This book is dedicated to -

-my family, for years they have hear me say, "I have to study."

- to my understanding wife, who I have missed companionship with so many evenings and week-ends. Our life together, can now be so much more active.

-and Aparna Munshi, a very special lady who will always be remembered for her very dignified demeanor and capacity for love.

Acknowledgments and thanks must go to-

-Raquel Brinkman, for helping with the business and money chapters.

-Aparna Munshi, for informing me, in such a nice way, that my first submission to her was totally incoherent.

-Rolland Walters , who made so many good suggestions for the Math and Algebra sections, that only a future edition can hold them all.

*-Carol Orim, for **loaning** me the first book that got me started studying money formulas.*

-Richard Morgret, for editing the complete book and for being a great critic, as well as a good friend.

Introduction

The formulas contained in this book are those that you will, or have encounter, through normal occurrences in your life. You now have the ability to have answers to these problems. By having the ability to figure out the answers on your own, you will then have more control and/or more knowledge to make the correct decisions that affect you.

The hardest part of math formulas is understanding what all of the symbols mean and the necessary steps to work the formula correctly. The goal of this book was to remove the symbols when ever possible and to show each step. In most examples only one element of an equation has been worked so that you can easily follow the steps taken. Please note that in ten point italic lettering, will be notes that aid you in working the problem. Whenever the word " calculate" is used that means to use a calculator. The only calculator used in the making of this book was a Texas Instrument Model 30 SLR, that is ten years old. The newer model TI-30Xa works as well. Both calculators cost about twenty dollars.

The amount of written words has been kept to a minimum, to make this text useful for easy and quick reference. In some cases correct grammar has given way to briefness and clarity.

The writer and publisher of this book, assumes no liability for the use or accuracy of the formulas or related information in this book. The results and usage is the sole responsibility of the user.

Recommended Calculators

Some sections of this book only require a typical calculator to aid in making calculations. However if you have any of the following calculators they will serve you well through out the book and with life in general. They all sell for less than $20.00.

Texas Instruments
TI 30 SLR, TI 30XIIS, TI 30Xa, TI36X

Casio
FX 115MS Plus, FX 260, FX 270W

Table of Contents

Table of Contents

Table of Contents

Table of Contents

Table of Contents

Copyright Law

All copyright laws apply to all of the material in this book. Instead of outlining the copyright laws, the following examples are given. However, there are situations that an example has not be given for. The fact of omission does not mean other situations are permissible.

Example- A teacher teaching a finance class, copies one of the Annuity formulas (there are six of them in this book). for all of the teachers students. This is permissible because it is for teaching purposes. However, if the teacher copies all six of the formulas, then this is of a complete section and it is a copyright infringement.

Example- The owner of a print shop, copies the page outlining paper cutting and attaches it to a paper cutter for the operators reference. This is permissible. However, if a copy was sent to another print shop, so that the additional print shop did not have to buy a book, then this is an infringement.

Example-A person is loaning their brother-in-law, money and the brother-in-law believes that the monthly payments are to high. It is permissible to fax or e-mail them a copy of the formula used from this book, proving the calculations were correct. However, if the person sent all of the multiplies on pages 9:57 to 9:61 it would be an infringement of copyright, because it then becomes a resource for another person.

Example- Posting even a part of a page to a web site is certainly an infringement of copyright. There is not a situation or example of when it would be permissible.

Multiply	By	To Obtain
Abamperes	10	Amperes
Abcoulbombs	2.998×10^{10}	Statcoulombs
Abfarads	$1.. \times 10^{9}$	Farads
Abfarads	$1. \times 10^{15}$	Microfarads
Abhenries	$1. \times 10^{-9}$	Henries
Abhenries	1×10^{-6}	Millihenries
Abohms	1×10^{-9}	Ohms
Abohms	1×10^{-15}	Megohms
Abvolts	$1. \times 10^{-8}$	Volts
Acres	10	Square chains (gunters)
Acres	160	Rods
Acres	100000	Square links
Acres	0.40047	Hectares
Acres	0.40047	Square hectometers
Acres	43560	Square feet
Acres	4047	Square meters
Acres	0.0016	Square miles
Acres	4840	Square yards
Acres	0.0015625	Section of Land
Acre feet	43560	Cubic feet
Acre feet	1233.48	Cubic meters
Acre feet	325900	Gallons
Amperes per sq. cm.	6.452	Amperes per sq. inch
Amperes per sq. cm.	10000	Amperes per sq. meter
Amperes per sq. inch	0.1550	Amperes per sq. cm.
Amperes per sq. inch	1550	Amperes per sq. meter
Amperes per sq. meter	0.0001	Amperes per sq. cm.
Amperes per meter	0.0006454	Amperes per inch
Ampere hours	3600	Coulombs
Ampere hours	0.03731	Faradays
Ampere turns	1.257	Gilberts
Ampere turns per cm.	2.540	Amp. turn per inch
Ampere turns per cm.	100	Amp turns per meter
Ampere turns per inch	0.3937	Amp turn per cm.
Ampere turns per inch	39.37	Amp turn per meter
Ampere turns per inch	0.4950	Gilberts per cm.
Ampere turns per meter	0.01	Amp turn per cm.
Ampere turns per meter	0.0254	Amp turns per inch
Ampere turns per meter	0.01257	Gilberts per cm.
Angstrom unit	3.937×10^{-9}	Inches
Angstrom unit	$1. \times 10^{-10}$	Meters
Angstrom unit	1×10^{-4}	Microns

CONVERSION LIST

Multiply	By	To Obtain
Angstroms	0.01	Centimeters
Ares	0.02471	Acres (U.S.)
Ares	119.6	Square yards
Ares	100	Square meters
Astronomical unit	1.495×10^8	Kilometers
Atmospheres	76	Centimeters
Atmospheres	29.92	Inches - mercury
Atmospheres	0.76	Meters of mercury at 0^0 C.
Atmospheres	1.0333	Kilograms per sq. cm.
Atmospheres	14.70	Pounds per square inch
Atmospheres	1.058	Tons per square foot
Atmospheres	0.007348	Tons per square inch
Atmospheres	33.9	Feet of water at 4^0 C.
Barrels (U.S.dry)	3.281	Bushels
Barrels (U.S.dry)	7056	Cubic inches
Barrels (U.S. liquid)	31.5	Gallons
Barrels - oil	42	Gallons
Bars	0.9869	Atmospheres
Bars	1,000,000	Dynes per sq. cm.
Bars	10,200	Kilograms per sq. meter
Bars	2089	Pounds per sq. foot
Bars	14.5	Pounds per sq. inch
Barye	1	Dynes per sq. cm.
Bolt (U.S. cloth)	36.576	Meters
Board feet	144	Cubic inches
British Thermal Units (BTU)	777.6	Foot Pounds
British Thermal Units (BTU)	.0003927	Horsepower hours
British Thermal Units	.0002928	Kilowatt - hours
Btu per minute	12.96	Foot pounds per second
Btu per minute	0.0236	Horsepower
Btu per minute	17.57	Watts
Btu	10.409	Liter per atmospheres
Btu	1.0550×10^{10}	Ergs
Btu	252	Grams of calories
Btu	1.055	Joules
Btu	3.929×10^{-4}	Horsepower
Bushels	1.2445	Cubic feet
Bushels	2150.4	Cubic inch
Bushels	0.03524	Cubic meters
Bushels	4 Pecks	
Bushels	64	Pints
Calories gram	3.9685×10^{-3}	BTU (mean)
Candle per sq cm.	3.146	Lamberts
Candle per sq. inch	0.4870	Lamberts

CONVERSION LIST

Multiply	By	To Obtain
Carat (diamond)	0.20	Grams
Centares	1	Square meters
Centigrade (degrees)	(^{0}C X 9/5) +32	Fahrenheit degrees
Centigrams	0.01	Grams
Centiliters	0.3382	Ounce (fluid U.S.)
Centiliters	0.6103	Cubic inch
Centiliters	0.01	Liters
Centimeters	0.3937	Inches
Centimeters	0.03281	Feet
Centimeters	0.01094	Yards
Centimeters	0.000006214	Miles
Centimeters	0.0001	Microns
Centimeters	1×10^8	Angstrom units
Centimeters	0.01	Meters
Centimeters	10	Millimeters
Centimeters Dynes	0.001020	Cn- grams
Centimeters Dynes	1.020×10^{-8}	Meter - kgs
Centimeters Dynes	7.376×10^{-8}	Pound feet
Centimeters Grams	980.7	Cm. dynes
Centimeters Grams	0.00001	Meter kgs.
Centimeter Grams	0.00007233	Pound feet
Centimeters - mercury	0.0132	Atmospheres
Centimeters - mercury	0.4460	Feet-water at 4^0 C
Centimeters- mercury	136.0	Kilograms/ square meter
Centimeters - mercury	27.85	Pounds per square foot
Centimeters - mercury	0.1934	Pounds per square inch
Centimeters per second	0.0328	Feet per second
Centimeters per second	0.036	Kilometers per hour
Centimeters per second	0.6	Meters per minute
Centimeters per second	0.0224	Miles per hour
Centimeters per second	0.0004	Miles per minute
Centipoise	0.01	Gr. per cm. a second
Centipoise	0.000672	Pounds per ft.. second
Cents	10	mills
Chains (gunters)	792	Inches
Chains (gunters)	20.12	Meters
Chains (gunters)	22	yards
Circular mils	0.000005067	Square cm.
Circular mils	0.7854	Square mils
Circular mils	7.854×10^{-7}	Square inches
Circumference	6.283	Radians
Cords	8	Cord feet
Cord feet	16	Cubic feet
Coulombs	2,998,000,000	Statcoulombs

CONVERSION LIST

Multiply	By	To Obtain
Coulombs	0.00001036	Faradays
Coulombs per square cm.	6.452	Coulombs per square inch
Coulombs per square cm.	10000	Coulombs per sq. meter
Coulombs per square inch	0.1550	Coulombs per square cm.
Coulombs per square inch	1550	Coulombs per sq meter
Coulombs per square meter	0.0001	Coulombs per sq cm.
Coulombs per square meter	0.0006452	Coulombs per sq inch
Cubic centimeters	.00003.531	Cubic feet
Cubic centimeters	0.0610	Cubic inches
Cubic centimeters	1 X 10 - 6	Cubic meters
Cubic centimeters	.000001379	Cubic yards
Cubic centimeters	.0002642	Gallons
Cubic centimeters	0.0010	Liters
Cubic centimeters	0.0021	Pints
Cubic centimeters	0.0011	Quarts
Cubic feet	1728	Cubic inches
Cubic feet	0.0283	Cubic meters
Cubic feet	28320	Cubic centimeters
Cubic feet	28.32	Liters
Cubic feet	29.92	Quarts
Cubic feet	7.4805	Gallons
Cubic feet	28.32	Liters
Cubic feet	59.84	Pints
Cubic feet	29.92	Quarts
Cubic feet per minute	0.1247	Gallons per second
Cubic feet per minute	0.4719	Liters per second
Cubic feet per minute	62.43	Pounds of water a minute
Cubic feet per second	448.831	Gallons per minute
Cubic inches	16.39	Cubic centimeters
Cubic inches	0.0005787	Cubic feet
Cubic inches	.0016387	Cubic meters
Cubic inches	.000021433	Cubic yards
Cubic inches	0.004329	Gallons
Cubic inches	0.0164	Liters
Cubic inches	0.0346	Pints
Cubic inches	0.0173	Quarts
Cubic meters	1,000,000	Cubic centimeters
Cubic meters	35.31	Cubic feet
Cubic meters	61023	Cubic inches
Cubic meters	1.308	Cubic yards
Cubic meters	264.2	Gallons
Cubic meters	1000	Liters
Cubic meters	2113	Pints
Cubic meters	1057	Quarts
Cubic yards	27	Cubic feet
Cubic yards	46.656	Cubic inches

CONVERSION LIST

Multiply	By	To Obtain
Cubic yards	0.7645	Cubic meters
Cubic yards	202	Gallons
Cubic yards	764.5	Liters
Cubic yards	1616	Pints
Cubic yards	807.9	Quarts
Cubic yards per minute	0.45	Cubic feet per second
Cubic yards per minute	3.367	Gallons per second
Cubic yards per minute	12.74	Liter per second
Daltons	1.650×10^{-24}	Grams
Decigrams	0.1	Grams
Deciliters	0.1	Liters
Decimeters	0.1	Meters
Degress (circle)	0.01111	Quadrants
Degrees (circle)	60	Minutes
Degrees (circle)	0.01745	Radians
Degrees (circle)	3600	Seconds
Degree per second	0.1667	Revolutions per minute
Degree per second	0.0028	Revolutions per second
Dekagrams (decagrams)	10	Grams
Dekaliters (decaliters)	10	Liters
Dekameters (decameters)	10	Meters
Drams	27.34	Grains
Drams	0.0625	Ounces
Drams	1.7718	Grams
Dynes per square cm.	0.01	Ergs per sq. millimeter
Dynes per square cm.	9.869×10^{-7}	Atmospheres
Dynes per square cm.	2.953×10^{-5}	Inches of mercury
Dynes per square cm.	4.015×10^{-4}	Inches of water at 4^{0} C.
Dynes	0.001020	Grams
Dynes	1.0×10^{-7}	Joules per cm.
Dynes	2.248×10^{-6}	Pounds
Ergs	9.486×10^{-11}	Btu
Ergs	1.0	Dyne centimeters
Ergs	7.376×10^{-8}	Foot pounds
Ergs	2.389×10^{-8}	Gram calories
Ergs	0.001020	Gram cms.
Ergs	3.7250×10^{-14}	Horsepower hours
Ergs	1.0×10^{-7}	Joules
Ergs	2.773×10^{-14}	Kilowatt hours
Ergs per second	5.668×10^{-9}	Btu per minute

CONVERSION LIST

Multiply	By	To Obtain
Ergs per second	4.426×10^{-6}	Foot lbs. per minute
Ergs per second	1.341×10^{-10}	Horsepower
Ergs per second	1×10^{-10}	Kilowatts
Farads	1,000,000	Microfarads
Faradays	26.8	Ampere hours
Faradays	96490	Coulombs
Fathoms	1.8288	Meters
Fathoms	6	Feet
Feet	30.48	Centimeters
Feet	12	Inches
Feet	0.3048	Meters
Feet	0.3333	Yards
Feet -water 4 degrees Cent.	0.8825	Inches of mercury
Feet of water	62.43	Pounds per square foot
Feet of water	0.0295	Atmospheres
Feet of water	304.8	kgs per sq meter
Feet per minute	0.5080	Centimeters per second
Feet per minute	0.0183	Kilometers per hour
Feet per minute	0.3048	Meters per minute
Feet per minute	0.0114	Miles per hour
Feet per second	30.48	Centimeter per second
Feet per second	1.097	Kilometers per hour
Feet per second	18.29	Meters per minute
Feet per second	0.6818	Miles per hour
Feet per second	0.0114	Miles per minute
Foot candles	10.764	Lumen per sq meter
Foot candles	10.764	Lux
Foot pounds	0.0013	British thermal units
Foot pounds	0.00000050505	Horse hours
Foot pounds	0.0000003766	Kilowatt hours
Foot pounds per minute	0.0167	Foot pounds per second
Foot pounds per minute	0.0030	Horsepower
Foot pounds per minute	0.0022597	Kilowatts
Furlongs	0.125	Miles
Furlongs	40	Rods
Furlongs	660	Feet
Furlongs	201.17	Meters
Gallons	3785	Cubic Centimeters
Gallons	0.1337	Cubic feet
Gallons	231	Cubic inches
Gallons	0.0038	Cubic meters
Gallons	3.785	Liters
Gallons	8	Pints
Gallons	4	Quarts
Gallons Imperial	1.2009	U.S. gallons

CONVERSION LIST

Multiply	By	To Obtain
Gallons U.S.	0.8327	Imperial gallons
Gallons of water	8.34	Pounds of water
Gallons per minute	0.002228	Cubic feet per second
Gallons per minute	0.06308	Liters per second
Gausses	6.452	Lines per square inch
Gausses	0.00000001	Webers per sq. cm.
Gausses	1.0-	Gilbert per cm.
Gilberts	0.7958	Ampere turns
Gilberts per cm.	0.7958	Ampere turns per cm.
Gilberts per cm	2.021	Ampere turns per inch
Gills (British)	142.07	Cubic cm.
Gills (U.S.)	118.295	Cubic cm.
Gills (U.S.)	0.183	Liters
Gills (U.S.)	0.25	Pints
Grade	0.01571	Radian
Grains	0.03657	Drams
Grains (Troy)	1.0	Grains (avdp)
Grains (Troy)	0.0648	Grams
Grains (Troy)	0.0020833	Ounces (avdp.)
Grains per U.S. gallon	17.118	Parts per million
Grains per U.S. gallon	142.86	Pounds per million gallons
Grains per imp. gallon	14.286	parts per million
Grams	980.7	Dynes
Grams	15.43	Grains
Grams	0.0353	Ounces
Grams	0.0322	Ounces (Troy)
Grams	0.002205	Pounds
Grams per cm.	0.0056	Pounds per inch
Grams per cubic centimeter	0.6243	Pounds per cubic feet
Grams per cubic centimeter	0.0361	Pounds per cubic inch
Grams per liter	58.17	Grains per gallon
Grams per liter	8.345	Pounds per 1,000 gal.
Grams per liter	0.062427	Pounds per cubic foot
Gram calories	0.0039683	Btu
Gram calories	41840000	Ergs
Gram calories	3.086	Foot pounds
Gram calories	1.5596×10^{-6}	Horsepower hours
Gram calories	1.162×10^{-6}	Kilowatt hours
Gram Centimeters	9.297×10^{-8}	Btu
Gram Centimeters	98.07	Ergs
Gram Centimeters	9.807×10^{-5}	Joules
Gram Centimeters	2.343×10^{-8}	Kilogram calories
Hand	10.16	Centimeters
Hand	4	Inches

CONVERSION LIST

Multiply	By	To Obtain
Hectares	2.471	Acres
Hectograms	100	Grams
Hectoliters	100	Liters
Hectometers	100	Meters
Hectowatts	100	Watts
Henries	1000	Millihenries
Hogsheads (British)	10.114	Cubic feet
Hogshead (U.S.)	8.42184	Cubic feet
Hogshead (U.S.)	63	Gallons
Horsepower	42.44	BTU per minute
Horsepower	33000	Foot pounds per minute
Horsepower	550	Foot pounds per second
Horsepower	1.014	Horsepower (Metric)
Horsepower	0.7457	Kilowatts
Horsepower	746	Watts
Horsepower (boiler)	33520	Btu per hour
Horsepower (boiler)	9.803	kilowatts
Horsepower hours	0.7457	Kilowatt hours
Horsepower hours	2547	Btu
Horsepower hours	2.6845×10^{13}	Ergs
Horsepower hours	1.98×10^{6}	Foot lbs.
Horsepower hours	641190	Gram calories
Horsepower hours	2684000	Joules
Horsepower hours	64.17	Kg. calories
Horsepower hours	273700	Kg. meters
Inches	25.4	Millimeters
Inches	2.540	Centimeters
Inches	0.0254	Meters
Inches	.0833333	Feet
Inches	.0277778	Yards
Inches	.00001578	Miles
Inches	2.54×10^{8}	Angstrom units
Inches	5.0505×10^{-3}	Rods
Inches of Mercury	0.033	Atmospheres
Inches of Mercury	345.3	Kilograms per sq. meter
Inches of Mercury	70.73	Pounds per square foot
Inches of water (4^{0}C.)	0.002458	Atmospheres
Inches of water (4^{0}C.)	0.5781	Ounces per square inch
Inches of water (4^{0}C.)	5.204	Pounds per square foot
Inches of water (4^{0}C.)	0.07355	Inches of mercury
Joules	0.0009486	Btu
Joules	10,000,000	Ergs
Joules	0.7736	Foot pounds

Multiply	By	To Obtain
Joules	2.389×10^{-4}	Kg. calories
Joules	0.102	Kg. meters
Joules	2.778×10^{-4}	Watt hours
Joules per centimeter	10200	Grams
Joules per centimeter	10,000,000	Dynes
Joules per centimeter	22.48	Pounds
Kilograms	980665	Dynes
Kilograms	2.205	Pounds
Kilograms	1000	Grams
Kilograms	0.09807	Joules per cm.
Kilograms	9.807	Joules per meter
Kilograms	0.0009842	Tons (long)
Kilograms	0.001102	Tons (short)
Kilograms	35.274	Ounces (avdp.)
Kilograms per cubic meter	1000	Grams per cubic cm.
Kilograms per cubic meter	0.06243	Pounds per cubic feet
Kilograms per meter	0.672	Pounds per foot
Kilograms per square cm.	980665	Dynes per square cm.
Kilograms per square cm.	0.9678	Atmospheres
Kilograms per square cm.	32.8	Feet of water
Kilograms per square cm.	28.96	Inches of Mercury
Kilograms per square cm.	2048	Pounds per square feet
Kilograms per square cm.	14.22	Pounds per square inch
Kilograms per square meter	0.009678	Atmospheres
Kilograms per square meter	9.807×10^{-5}	Bars
Kilograms per square meter	0.003281	Feet of water
Kilograms per square meter	0.002896	Inches of mercury
Kilograms per square meter	0.2048	Pound per square feet
Kilograms per square meter	0.001422	Pounds per square inch
Kilograms per square meter	98.0665	Dynes per square cm.
Kilogram calories	3.968	Btu
Kilogram calories	0.003086	Foot pounds
Kilogram calories	0.001558	Horsepower hours
Kilogram calories	0.004183	Joules
Kilogram calories	0.001163	Kilowatt hours
Kiloliters	0.001	Liters
Kiloliters	1.308	Cubic yards
Kiloliters	0.35316	Cubic feet
Kiloliters	0.026418	Gallons (U.S.)
Kilometers	1000000	Millimeters
Kilometers	100000	Centimeters
Kilometers	1000	Meters
Kilometers	39370	Inches
Kilometer	3281	Feet

CONVERSION LIST

Multiply	By	To Obtain
Kilometers	1094	Yards
Kilometers	0.6214	Miles
Kilometers	0.00000000001057	Light years
Kilometers	1.0570×10^{-11}	Light years
Kilometers per hour	54.68	Feet per minute
Kilometers per hour	0.6214	Miles per hour
Kilometers per hour	16.67	Meter per minute
Kilometers per hour	0.5396	Knots
Kilowatts	56.82	BTU per minute
Kilowatts	44253.7	Foot pounds per minute
Kilowatts	737.6	Foot pounds per second
Kilowatts	1.341	Horsepower
Kilowatt hours	3410	BTU
Kilowatt hours	3.6×10^{13}	Ergs
Kilowatt hours	3.6×10^{6}	Joules
Kilowatt hours	2655000	Foot pounds
Kilowatt hours	1.341	Horsepower hours
Knots	6080	Feet per hour
Knots	1.8532	Kilometers per hour
Knots	1.0	Nautical miles per hour
Knots	1.151	Miles per hour
Lambert	0.3183	Candle per square cm.
Lambert	2.054	Candle per square inch
League	3.0	Miles
Light years	9,460,500,000,000	Kilometers
Light years	9.4605×10^{12}	Kilometer
Light years	0.3066	Parsecs
Light years	5,900,000,000,000	Miles
Light years	5.9×10^{12}	Miles
Lines per square cm.	1.0	Gausse
Lines per square inch	0.155	gausse
Lines per square inch	1.55×10^{-9}	Webers per square cm.
Lines per square inch	1.0×10^{-8}	Webers per square inch
Links (engineers)	12	Inches
Links (surveyors)	7.92	Inches
Liters	61.02	Cubic inches
Liters	0.353	Cubic feet
Liters	0.001308	Cubic yards
Liters	0.2642	Gallons
Liters	2.113	Pints
Liters	1.057	Quarts
Liters	0.001	Cubic centimeters
Liters	0.0010	Cubic meters

CONVERSION LIST

Multiply	By	To Obtain
Lumen	0.07958	Spherical candle power
Lumen per square feet	1.0	Foot candles
Lumen per square feet	10.76	Lumen square meters
Lux	0.0929	Foot candles
Maxwells	0.001	Kilolines
Maxwells	1.0×10^{8}	Webers
Megalines	1.0×10^{6}	Maxwells
Megohms	1.0×10^{12}	Microhms
Megohms	1.0×10^{6}	Ohms
Megohms per cubic cm.	0.001	Abmhos per cubic cm.
Megohms per cubic cm.	2.54	Megohms per cubic inch
Megohms per cubic cm.	0.1662	Megohms per mil ft.
Meters	39.37	Inches
Meters	3.281	Feet
Meters	1.094	Yards
Meters	0.0005396	Miles (nautical)
Meters	0.0006214	Miles (statute)
Meters	0.001	Millimeters
Meters	0.01	Centimeters
Meters	0.001	Kilometers
Meters	1.0×10^{10}	Angstrom units
Meters	0.54681	Fathoms
Meters per minute	3.281	Feet per minute
Meters per minute	0.06	Kilometers per hour
Meters per minute	0.03733333	Miles per hour
Meters per second	196.8	Feet per minute
Meters per second	3.281	Feet per second
Meters per second	3.6	Kilometers per hour
Meters per second	0.03728	Miles per minute
Microfarads	1.0×10^{-15}	Abfarads
Microfarads	1.0×10^{-6}	Farads
Microfarads	900000	Statfarads
Micrograms	0.000001	Grams
Microhms	1000	Abohms
Microhms	1.0×10^{-12}	Megohms
Microhms	1.0×10^{-6}	Ohms
Microliters	1.0×10^{-6}	Liters
Micromicrons	1.0×10^{-12}	Meters
Microns (micrometers)	1,000,000	Meters
Microns (micrometers)	100,000	Angstroms

CONVERSION LIST

Multiply	By	To Obtain
Miles (nautical)	1853	Meters
Miles (nautical)	1.853	Kilometers
Miles (nautical)	6076	Feet
Miles (nautical)	2025.4	Yards
Miles (statute)	160900	Centimeters
Miles (statute)	1609	Meters
Miles (statute)	1.609	Kilometers
Miles (statute)	63360	Inches
Miles (statute)	5280	Feet
Miles (statute)	1760	Yards
Miles (statute)	1.69×10^{-13}	Light years
Miles per hour	44.7	Centimeters per second
Miles per hour	88	Feet per minute
Miles per hour	1.467	Feet per second
Miles per hour	1.609	Kilometer per hour
Miles per hour	0.8690	Knots
Miles per hour	26.82	Meter per minute
Miles per minute	2682	Centimeters per second
Miles per minute	88	Feet per second
Miles per minute	60	Miles per hour
Miles per minute	1.609	Kilometers per minute
Milliliter	0.001	Kiloliter
Millimicrons	1.0×10^{-9}	Meters
Milligrams	0.015432	Grains
Milligrams	0.001	Grams
Milligrams per liter	1.0	Parts per million
Millihenries	0.001	Henries
Milliliters	0.001	Liters
Millimeters	0.1	Centimeters
Millimeters	0.001	meters
Millimeters	0.03937	Inches
Millimeters	0.003281	Feet
Mills	0.00254	Centimeters
Mills	8.333×10^{-5}	Feet
Mills	0.001	Inches
Mills	0.10	Cents
Miner's inch	1.5	Cubic feet per minute
Minims (British)	5.9192×10^{-2}	Cubic cm.
Minims (U.S.fluid)	6.1612×10^{-2}	Cubic cm.
Minutes (angle)	0.01667	Degrees
Minutes (angle)	0.00018752	Quadrants
Minutes (angle)	0.0002909	Radians
Minutes (angle)	60	Seconds
Myriagrams	10	Kilograms
Myriameters	10	Kilometers

CONVERSION LIST

Multiply	By	To Obtain
Nails	2.25	Inches
Newtons	100000	Dynes
Ohms	1.0×10^{-6}	Megohms
Ohms	1,000,000	Microhms
Ounces	8.0	Drams
Ounces	437.5	Grains
Ounces	28.349	Grams
Ounces	0.0625	Pounds
Ounces	0.9115	Ounces (Troy)
Ounces (fluid)	1.805	Cubic inches
Ounces (fluid)	0.02957	Liters
Ounces (troy)	480	Grains
Ounces (troy)	31.103	Grams
Ounces (troy)	1.097	Ounces (avdp.)
Ounces (troy)	20	Pennyweights
Ounces (troy)	0.08333	Pounds (troy)
Ounces per square inch	430.9	Dynes per square cm.
Ounces per square inch	0.0625	Pounds per square inch
Ounces	16	Drams
Ounces	437.5	Grains
Ounces	0.0625	Pounds
Ounces	0.9115	Ounces (Troy)
Ounces	0.000028349	Tons (Metric)
Ounces (Troy)	1.0-971	Ounces (Avoirdupois)
Ounces	1.805	Cubic inches
Ounces	0.0625	Fluid Pint
Ounces	0.03125	Fluid Quart
Ounces	0.0296	Liters
Pace	30	Inches
Parsec	1.9×10^{13}	miles
Parsecs	30,860,000,000,000	Kilometers
Parsecs	3.086×10^{13}	Kilometers
Parsecs	3.2617	Light years
Parts per million	0.0584	Grains per U.S. gallon
Parts per million	8.345	Pounds per million gallons
Pecks (British)	554.6	Cubic inches
Pecks (British)	9.0919	Liters
Pecks	0.25	Bushels
Pecks	537.6	Cubic inches
Pecks	8.8096	Liters
Pecks	8	Quarts
Pennyweights (troy)	24	Grains
Pennyweights (troy)	0.5	Ounces (troy)
Pennyweights (troy)	1.555	Grams
Pennyweights (troy)	0.0041667	Pounds (troy)

CONVERSION LIST

Multiply	By	To Obtain
Pints (dry)	33.6	Cubic inches
Pints (dry)	0.015625	Bushels
Pints (dry)	0.55059	Liters
Pints (dry)	0.5	Quarts
Pints	16	Fluid Ounces
Pints (liquid)	0.01671	Cubic feet
Pints (liquid)	28.87	Cubic inches
Pints (liquid)	0.0006189	Cubic yards
Pints (liquid)	0.125	Gallons
Pints (liquid)	473.2	Cubic cm.
Pints (liquid)	0.0004732	Cubic meters
Pints (liquid)	0.4732	Liters
Pints (liquid)	0.5	Quarts
Planck's quantum	6.547×10^{-27}	Erg seconds
Poise	1	Grams per cm. sec.
Poundals	13826	Dynes
Poundals	14.1	Grams
Poundals	0.001383	Joules per cm.
Poundals	0.1383	Joules per meter
Poundals	0.0141	Kilograms
Poundals	0.03108	Pounds
Pounds	16	Ounces
Pounds	256	Drams
Pounds	7000	Grains
Pounds	453.59	Grams
Pounds	0.04448	Joules per cm.
Pounds	0.4536	Kilograms
Pounds	32.17	Poundals
Pounds	0.0005	Tons (Short)
Pounds	0.0004464	Tons (Long, British)
Pounds	1.2153	Pounds (Troy)
Pounds (troy)	5760	Grains
Pounds (troy)	373.24	Grams
Pounds (troy)	13.166	Ounces (avdp.)
Pounds (troy)	12	Ounces (troy)
Pounds (troy)	240	Pennyweights (troy)
Pounds (troy)	0.82286	Pounds (avdp.)
Pounds (troy)	3.6735×10^{-4}	Tons (long)
Pounds (troy)	3.7324×10^{-4}	Tons (metric)
Pounds (troy)	4.1143×10^{-4}	Tons (short)
Pounds of water	0.01602	Cubic feet
Pounds of water	27.68	Cubic inches
Pounds of water	0.1198	Gallons
Pound feet	13825	Cm. grams
Pounds per inch	178.6	Grams per centimeter

CONVERSION LIST

Multiply	By	To Obtain
Pounds per cubic inch	1728	Pounds per cubic foot
Pounds per square foot	1488	Kilograms per meter
Pounds per square foot	4.882	Kilograms per sq. meter
Pounds per square foot	4.725 X10^{-4}	Atmospheres
Pounds per square foot	0.01602	Feet of water
Pounds per square foot	0.01414	Inches of mercury
Pounds per square foot	0.006944	pounds per square inch
Pounds per square inch	144	pounds per square ft.
Pounds per square inch	0.0680	Atmospheres
Pounds per square inch	2.036	Inches of mercury
Quadrants (circle)	1.571	Radians
Quadrants (circle)	324000	Seconds
Quadrants (circle)	54000	Minutes
Quadrants (circle)	90	Degrees
Quarts (dry)	67.2	Cubic inches
Quarts (liquid)	57.75	Cubic inches
Quarts (liquid)	0.03342	Cubic feet
Quarts (liquid)	0.25	Gallons
Quarts (liquid)	946.4	Cubic cm.
Quarts (liquid)	0.0009464	Cubic meters
Quarts (liquid)	0.9463	Liters
Quintal (metric)	100	Kilograms
Quintal (metric)	220.46	Pounds
Radians	57.3	Degrees
Radians	3438	Minutes
Radians	0.637	Quadrants
Radians per second	9.549	Revolutions per minute
Reams (paper)	500	Sheets
Revolutions per second	360	Degrees per second
Revolutions per second	6.283	Radians per second
Revolutions per second	60	Revolutions per minute
Rods	0.25	Chains (gunthers)
Rods	5.029	Meters
Rods (surveyor's)	5.5	Yards
Rods	16.5	Feet
Rods	198	Inches
Rods	0.003125	Miles
Rope	20	Feet
Scruples	20	Grains
Seconds (circle)	0.01667	Minutes
Seconds "	0.0002778	Degrees
Seconds "	0.000003087	Quadrants
Seconds "	0.0000048481	Radians
Seconds (time)	10^{-15}	Femtosecond
Seconds "	10^{-12}	Picosecond
Seconds "	10^{-9}	Nanosecond

CONVERSION LIST

Multiply	By	To Obtain
Seconds (time)	10^{-6}	Microsecond
Seconds "	10^{-3}	Millisecond
Section of Land	640	Acres
Slugs	14.59	Kilograms
Slugs	32.17	Pounds
Sphere (solid angle)	12.57	Steradians
Square centimeters	0.0011	Square feet
Square centimeters	0.1550	Square inches
Square centimeters	0.0001	Square meters
Square centimeters	100	Square millimeters
Square degrees	0.00030462	Steradians
Square feet	0.0022957	Acres
Square feet	144	Square inches
Square feet	0.1111	Square yards
Square feet	0.00000003587	Square miles
Square feet	3.5870×10^{-8}	Square miles
Square feet	t929	Square centimeters
Square feet	0.0929	Square meters
Square inches	6.452	Square centimeters
Square inches	0.0069	Square feet
Square inches	1273000	Circular mils
Square kilometers	247.1	Acres
Square kilometers	10,764,000	Square feet
Square kilometers	1,000,000	Square meters
Square kilometers	0.3861	Square miles
Square kilometers	1,196,000	Square yards
Square meters	10.76	Square feet
Square meters	1.1960	Square yards
Square meters	0.0002471	Acres
Square meters	10,000	Square cms.
Square meters	1,000,000	Square millimeters
Square miles	640	Acres
Square miles	2.590	Square kilometers
Square miles	2,590,000	Square meters
Square miles	3,097,600	Square yards
Square millimeters	0.01	Square centimeters
Square millimeters	0.0016	Square inches
Square mils	1.273	Circular mils
Square mils	0.000006452	Square cms.
Square mils	0.000001	Square inches
Square yards	1296	Square inches
Square yards	9	Square feet
Square yards	0.00000032283	Square miles
Square yards	3.2283×10^{-7}	Square miles
Square yards	0.8361	Square meters

CONVERSION LIST

Multiply	By	To Obtain
Square yards	0.0002066	Acres
Steradians	0.07958	Spheres
Steradians	0.1592	Hemispheres
Steradians	0.6366	Spherical right angles
Steres	999.973	Liters
Tons (long)	1016	Kilograms
Tons (long)	2240	Pounds
Tons (long)	1.12	Tons (short)
Tons (Metric)	1000	Kilograms
Tons (Metric)	2205	Pounds
Tons (Short) U.S.	2000	Pounds
Tons (Short) U.S.	0.89286	Tons (Long)
Tons (Short) U.S.	0.9072	Tons (Metric)
Tons (short)	907.18	Kilograms
Volt per inch	39370000	Abvolt per cm.
Volt per inch	0.3937	Volt per cm.
Volt	100,000,000	Abvolts
Watts	0.0586	Btu per minute
Watts	0.7377	Foot pounds per second
Watts	0.0013	Horsepower
Watts	0.001	Kilowatts
Watt hours	3.4144	Btu
Watt hours	2655	Foot pounds
Watt hours	0.00134	Horsepower hours
Watt hours	0.001	Kilowatt hours
Watt hours	3.6×10^{10}	Ergs
Watt hours	860.5	Gram calories
Webers	1.0×10^{8}	Maxwells
Webers	1.0×10^{5}	Kilolines
Webers per square inch	1.55×10^{7}	Gausse
Webers per square inch	1.0×10^{8}	Lines per square inch
Webers per square inch	0.155	Webers per square cm.
Webers per square inch	155	Webers per square meter
Webers per square meter	100,000	Gausse
Webers per square meter	64520	Lines per square inch
Webers per square meter	0.0001	Webers per square cm.
Yards	91.44	Centimeters
Yards	0.0009144	Kilometers
Yards	0.9144	Meters
Yards	49340	Miles (nautical)
Yards	56820	Miles (statute)
Yards	3	Feet
Yards	36	Inches

Metric
Conversion Table

Length

Unit	Abbreviation	Number of Meters	U.S. Equivalent
gigameter	G	1,000,000,000	621,400 miles
megameter	M	1,000,000	621.4 miles
kilometer	km	1,000	0.6214 mile
hectometer	hm	100	328.08 feet
dekameter	dam (decameter)	10	32.81 feet
meter	m	1	39.37 inches
decimeter	dm	0.1	3.94 inches
centimeter	cm	0.01	0.39 inch
millimeter	mm	0.001	0.039 inch
micrometer	µm (µ micron)	0.000001	0.000039 inch

1000 micrometers	=	1 millimeter
10 millimeters	=	1 centimeter
10 centimeters	=	1 decimeter
10 decimeters	=	1 meter
10 meters	=	1 dekameter
10 dekameters	=	1 hectometer
10 hectometers	=	1 kilometer
1,000 kilometers	=	1 megameter
1,000 mega meters	=	1 gigameter

Liquid

Unit	Abbreviation	Number of Liters	U.S. Equivalent
kilometer	kl	1,000	1.31 cubic yards
hectoliter	hl	100	3.53 cubic feet
dekaliter	dal (decaliter)	10	0.35 cubic foot
liter	l $_3$	1	61.02 cubic inches
cubic decimeter	dm	1	61.02 cubic inches
deciliter	dl	0.1	6.1 cubic inches
centiliter	cl	0.01	0.61 cubic inch
milliliter	ml	0.001	0.061 cubic inch
microliter	µl	0.000001	0.000061 cu. in

Weight

Unit	Abbreviation	Number of Grams	U.S. Equivalent
metric ton	t	1,000,000	1.102 short tons
kilograms	kg	1,000	2.2046 pounds
hectograms	hg	100	3.527 ounces
dekagrams	dag (decagram)	10	0.353 ounces
gram	g	1	0.035 ounce
decigram	dg	0.1	1.543 grains
centigram	cg	0.01	0.154 grain
milligram	mg	0.001	0.0154 grain
microgram	µg	0.000001	0.000015 grain

Metric
Conversion Table

Length and Area

To change from:	To:	Multiply by:
inches	millimeters	25.4
inches	centimeters	2.540005
feet	centimeters	30.48006
yards	meters	0.914402
miles	kilometers	1.60935
millimeters	inches	0.03937
centimeters	inches	0.3937
meters	feet	3.25083
kilometers	miles	0.6211372
square inches	square centimeters	6.461626
square feet	square meters	0.0929034
square yards	square meters	0.8361307
acres	ares	40.46973
square miles	square kilometers	2.589998
square centimeters	square inches	0.155
square meters	square feet	10.76387
square kilometers	square miles	0.3861006

Volumes

To change from:	To:	Multiply by:
cubic inches	cubic centimeter (cc)	16.38716
cubic feet	cubic meter	0.028317
cubic yard	cubic meter	0.764559
cubic centimeter	cubic inch	0.0610234
cubic meter	cubic foot	35.3144
cubic meter	cubic yard	1.30794

Metric
Conversion Table
Liquid Capacity

To change from:	To:	Multiply by:
ounce	cubic centimeter	29.57
pint	liter	0.473167
quart	liter	0.946333
gallon	liter	3.785332
liter	fluid ounce	33.8147
liter	quart	1.05671
liter	gallon	0.264178
cubic centimeter	ounce	0.033815

Dry Capacity

To change from:	To:	Multiply by:
pint	liter	0.550599
quart	liter	1.101198
peck	liter	8.80958
bushes	liter	35.2383
liter	pint	1.81620
liter	quart (qts.)	0.908102
deciliter	peck	1.13513
hectoliter	bushel (bu)	2.83782

Avoirdupois Weights

To change from:	To:	Multiply by:
grain	gram	0.0647989
ounce	gram	28.349527
pound	kilogram (kg)	0.453924
ton	kilogram	907.18486
gram	grain	15.43235639
kilogram	pound (lbs.)	2.2046223
metric ton	pound	2204.6223

Equivalent Lists
by
Subject Matter

Area

144 square inches	= 1 square foot
9 square feet	= 1 square yard
30.25 square yards	= 1 square rod
1 rod	= 10,890 square feet
1 rod	= 40 square rods
1 acre	= 43.560 square feet
160 square rods	= 1 acre
640 acres	= 1 square mile
640 Acres	= 1 section of land

Avoirdupois Weight

1 dram	= 27.34375 grains
16 drams (dr.)	= 1 ounce
7000 grams (gr)	= 1 pound
16 ounces (oz.)	= 1 pound
100 pounds	= 1 hundredweight (cwt.)
2000 pounds	= 1 ton (T)
2240 pounds	= 1 long ton

British Liquid

1 British imperial quart	= 1.20095 U.S. liquid quarts
4 British imperial quarts	= 1 British imperial gallon
36 British imperial gallons	= 1 British Barrel
1 British dry quart	= 1.0329 U.S. dry quarts
14 pounds	= 1 stone
112 pounds	= 1 hundredweight
20 grains	= 1 scruple
3 scruples	= 1 dram
3 drams	= 1 ounce

Equivalent Lists
by
Subject Matter

Circle and Arc

60 seconds	= 1 minute
60 minutes	= 1 degree
30 degrees	= 1 sign
90 degrees	= 1 quadrant
4 quadrants	= 360 degrees
360 degrees	= 1 circle
1 minute	= 60 seconds of arc
1 degree	= 60 minutes of arc

Cubic

1728 cubic inches	= 1 cubic foot
27 cubic feet	= 1 cubic yard
128 cubic feet	= 1 cord
1 cubic inch	= 16.387064 cubic centimeters

Dry Measurement

2 pints	= 1 quart
8 quarts	= 1 peck (pk.)
4 pecks	= 1 bushel (bu.)
1 bushel	= 2150.42 cubic inches
1 cubic foot	= 1,728 cubic inches
1 cubic yard	= 27 cubic feet
1 pint	= 33.6003125 cubic inches
1 quart	= 67.200625 cubic inches
1 gallon	= 268.8025 cubic inches
1 peck	= 537.605 cubic inches
1 bushel	= 2,150.42 cubic inches
1 barrel	= 7,056 cubic inches
1 cord foot (wood)	= 16 cubic feet
1 cord (wood)	= 128 cubic feet
1 freight ton	= 40 cubic feet
1 register ton	= 100 cubic feet

Equivalent Lists

by
Subject Matter

Distance

1 nail (cloth)	= 2.25 inches
1 palm	= 3 inches
1 hand	= 4 inches
1 span	= 6 inches
1 quarter (cloth)	= 9 inches
1 pace	= 30 inches
inches (in) (″)	= 1 foot (ft.) (′)
3 feet (ft)′	= 1 yard (yd.)
16.5 feet	= 1 rod (rd.)
320 rods	= 1 mile
1 fathom	= 6 feet
1 rod	= 16.5 feet
1 furlong	= 660 feet
1760 yards	= 1 mile
5280 feet	= 1 mile
1 nautical mile	= 6076.10333 ft.
1 nautical mile	= 1.15078 miles
1 nautical mile	= 1852 meters
1 fathom	= 6 feet
1 cable	= 120 fathoms
1 cable	= 720 feet

Liquid

80 drops	= 1 tablespoon
3 teaspoon	= 1 tablespoon
2 tablespoon	= 1 fluid ounce
3 tablespoon	= 1 jigger
4 tablespoon	= 1/4 cup
16 tablespoons	= 1 cup
2 cups	= 1 pint
1 cup	= 8 fluid ounces
1 fluid dram	= 60 minims
1 teaspoon	= 80 minims
1 tablespoon	= 240 minims
1 fluid ounce	= 2 tablespoons
1 gill	= 4 fluid ounces

Equivalent Lists
by
Subject Matter

Liquid (continued)

4 quills (qi.)	= 1 pint (pt.)
1 pint	= 16 fluid ounces (oz.)
2 pints	= 1 quart (qt.)
4 quarts	= 1 gallon (gal.)
1 gallon	= 128 fluid ounces
231 cubic inches	= 1 gallon
31.5 gallons	= 1 barrel (bbl.)
1 petroleum barrel	= 42 gallons

Pressure

1 cubic foot of water	= 62.425 pounds
1 cubic feet per second	= 448.83 gallons per minute
1 pound per square inch	= 27.7 inches of water pressure
1 pound per square inch	= 2.31 feet of water pressure
1 pound per square inch	= 2.036 inches of mercury pressure
1 kilowatt-hour	= 2.655×10^{6} foot pounds

Time

Femtosecond	fs	10^{-15} second
Picosecond	ps	10^{-12} second
Nanosecond	ns	10^{-9} second
Microsecond	μs	10^{-6} second
Millisecond	ms	10^{-3} second

60 seconds = 1 minute; 60 minutes = 1 hour; 24 hours = 1 day; 7 days = 1 week; 1 fortnight = 2 weeks; 365 days = 1 common year; 366 days = 1 leap year (every 4 years); 1 lunar month = 29 days, 12 hours, 44 minutes; 1 olympiad = 4 years; 1 sexennial = 6 years; 1 septennial = 7 years; 1 octennial = 8 years; novennial = 9 years; 1 decade = 10 years; 1 duode-cennial = 12 years; quindecennial = 15 years; 1 tricennial = 30 years; 1 semicentennial = 50 years; 1 century = 100 years; 1 sesquicentennial = 150 years; 1 bicentennial = 200 years; 1 millennium = 1,000 years;

Equivalent Lists

by
Subject Matter

Troy Weight

1 scruple	= 20 grains
24 grains	= 1 pennyweight (pwt)
1 dram	= 60 grains
1 ounce	= 480 grains
1 ounce	= 8 drams
20 pennyweights	= 1 ounce (oz.)
12 ounces	= 1 pound
5760 grains	= 1 pound
3.168 grains	= 1 carat

Velocity/ Speed

1 mile per hour = 88.028 feet per minute
1 mile per hour = 1.467 feet per second
Light = 186,291 miles per second
Sound in Air = 742 miles per hour
Sound in Water 32 degrees F = 3,246 mph
Jet- typical passenger = 600 mph
Jet - Concorde = 1,300 mph
Guns
   High velocity 22/250 Caliber = 3,850 fps
   Low velocity 38 Caliber = 675 fps
   Average = 2,940 feet per second
Mach 1 = 742 mph
Rockets = 17,500 mph

Equivalent Lists
Astronomical Numbers

Light Year	5,880,000,000,000 miles
Velocity of light	186,282 miles per second
Earth to Moon average distance	238,860 miles
Radius of Earth at the Equator	3,963.34 statute miles
Radius of Earth at N. or S. Poles	3,949.99 statute miles
Earth's average velocity in orbit	18.5 miles per second
Earth's velocity around the Sun	67,000 miles per hour
Earth's orbit around the Sun	584,050,000 miles

Earth's orbit time 365 days, 5 hours, 48 minutes, 45.51 seconds
Earth's rotation   23 hours, 56 minutes, 4.09 seconds

Earth's weight	6.6 sextillion tons
Earth's Surface	196,949,970 square miles
Nearest Star Alpha Centauri	4 light years away
Farthest Star Canopus	650 light years away

Distance from the Sun

Mercury	36	million miles
Venus	67.24	" "
Earth's	92.9	
Mars	141.71	
Jupiter	483.88	
Saturn	887.14	
Uranus	1,783.98	
Neptune	2,796.46	
Pluto	3,666	
Suns temperature core	20 million degrees Celsius	
outer surface	6,000 degrees Celsius	

Diameter

Sun	865,400 miles
Mercury	3,032
Venus	7,519
Earth	7,962
Mars	4,194
Jupiter	88,736
Saturn	74,978
Uranus	32,193
Neptune	30,775
Pluto	1,423

Addition

2+2=4	$\begin{array}{r} 2 \\ +\ 2 \\ \hline 4 \end{array}$	13+5 = 18 $\begin{array}{r} 13 \\ +\ 5 \\ \hline 18 \end{array}$	53+5 = 58 $\begin{array}{r} 53 \\ +\ 5 \\ \hline 58 \end{array}$

When adding large numbers, each column can hold only nine units. When you get ten of that unit, you carry it over to the next column.

45+7 = 52 $\begin{array}{r} ^145 \\ +\ 7 \\ \hline 52 \end{array}$ In this problem 5 + 7 equals 12. The 2 is put below the 7 and the 1 is carried over to the 4. 4 + 1 equals 5.

96+68+75 = 239 $\begin{array}{r} ^196 \\ +\ 68 \\ +\ 75 \\ \hline 239 \end{array}$ 6 + 8 + 5 = 19. Write the 9 in the answer. Carry the 1 over to the next column. Add 9 and 6 and 7 equaling 22. With the 1 carried over from the previous column write 23 next to the 9 making the total 239.

When adding numbers with decimal points, you line up the numbers to the decimal point.

$$14.7 + 3.95 + 695.467 + 29.2$$

$$\begin{array}{r} ^{1}2\ 2\ ^1 \\ 14.7 \\ 3.95 \\ 695.467 \\ \underline{29.2\ \ \ } \\ 743.317 \end{array}$$

Addition

Adding columns of large numbers

Total each column then add these together.

```
   2 5 8 . 1 3
   2 4 5 . 1 0
 1 9 7 3 . 3 3
 4 5 5 3 . 9 9
 3 2 1 1 . 6 7
 3 1 1 2 . 3 3
   2 4 6 . 1 2
   2 1 0 . 0 0
+  5 6 1 . 3 3
```

.3 0	total of first column on right
2 .7	total of second column from right
2 9	total of third column from right
3 4	total of forth column from right
3 0	total of fifth column from right
1 1	total of sixth column from right

1 4 3 7 2 . 0 0

Transposing Errors - Divisible by 9

When writing long columns of numbers or adding them, it is possible to transpose them. Subtracting the lowest total from the highest total and then dividing the answer by 9, can help detect if two numbers were transposed. If the results are divisible by 9, without a remainder then there is a good probability that the difference is due to a transposing error of two numbers.

Example -

```
 315          315        2252
 555          555      − 2072        180
 685          865         180        ───  = 20
 517          517                     9
────         ────
2072         2252
```

Addition

Adding Random Numbers in the Cents Column

You can average the cents column only if you have truly random numbers, such as a grocery bill or an electric bill that is based on the actual amount used.

230.23	Total only the whole dollar
232.09	amounts and not the cents
130.93	columns. The total will be 2,834.
239.47	Then total the number of entries
216.97	which is 12 . Divide this total
200.37	by 2 equaling 6. Add this amount
235.45	to 2,834 = $2,840.00. Please
266.42	note that this amount is
359.94	0.03 more than the actual
239.23	amount of $2,839.97. In most
253.69	cases by averaging the cents
235.18	column this total will vary less
2,839.97	than $1.00. from the actual total.

Subtraction

$$7 - 2 = 5 \qquad \begin{array}{r} 7 \\ - \ 2 \\ \hline 5 \end{array} \qquad 18 - 5 = 13 \qquad \begin{array}{r} 18 \\ - \ 5 \\ \hline 13 \end{array} \qquad 37 - 6 = 31 \qquad \begin{array}{r} 37 \\ - \ 6 \\ \hline 31 \end{array}$$

$$42 - 9 = 33 \qquad \begin{array}{r} 42 \\ - \ 9 \\ \hline 33 \end{array} \qquad \begin{array}{r} ^{3}4 \ ^{1}2 \\ - \ \ 9 \\ \hline 3 \ 3 \end{array}$$

In this problem the 9 is greater than 2, so you take 10 units from 4 making the 2 equal 12. 9 from 12 is 3. The 4 was changed to 3.

$$86 - 57 = 29 \qquad \begin{array}{r} 86 \\ - \ 57 \\ \hline 29 \end{array} \qquad \begin{array}{r} ^{7}8 \ ^{1}6 \\ - \ 5 \ 7 \\ \hline 2 \ 9 \end{array}$$

The 6 is less than 7 so 10 units are taken from the 8 making 6 equal 16. 7 from 16 is 9. The 8 became 7 when 10 units were taken from it. 5 from 7 equals 2.

Multiplication

When you want double of an item you multiply it by 2 .

$3 \times 2 = 6$ 3
           X 2
           6

$5 \times 2 = 10$ 5
           X 2
           10

$14 \times 2 = 28$ 14
           X 2
           28

$135 \times 2 = 270$ $1^{1}3\,5$
           X   2
           2 7 0

2X5=10 , put the 0 down and carry the 1 over to next column and add 1 to that column. 2X3 is 6 and with the 1 from the column on the right you add it to the 6 and get 7

When you want to have 3,4,5,6,7,8 or 9 times the amount, you do as you did above.

$5 \times 6 = 30$ 5
           X 6
           30

$18 \times 6 = 108$ $^{4}18$
           X 6
           108

$362 \times 7 = 2534$ $^{4}3\,^{1}6\,2$
           X   7
           2 5 3 4

In the above problem 8 X 6 is 48. The 8 is put in the units column and the 4 was carried over to next column and added to 1 X 6 equaling 10

In the above problem 2 X 7 is 14. The 4 is put in the units column and the 1 was carried over to the next column and added to 6 X 7 equaling 43. the 4 from 43 was carried over to the next column to be added to 3 X 7 equaling 25

$75 \times 32 = 2400$ $^{1}75$
           X 32
           150
           225
           2400

In this problem 5X2 is 10, put the 0 down below and carry the 1 over to the 7. 2X7 is 14, add the 1 and put 15 down below. 3X5 is 15, put the 5 directly below the same column the 3 is in and carry the 1 over to the 7, 3X7 is 21, now add the 1 and put 22 down below. Now add 150 and 225 as they are in their respective columns equaling 2400.

Multiplication

MULTIPLICATION TABLE

	1	2	3	4	5	6	7	8	9
1	1	2	3	4	5	6	7	8	9
2	2	4	6	8	10	12	14	16	18
3	3	6	9	12	15	18	21	24	27
4	4	8	12	16	20	24	28	32	36
5	5	10	15	20	25	30	35	40	45
6	6	12	18	24	30	36	42	48	54
7	7	14	21	28	35	42	49	56	63
8	8	16	24	32	40	48	56	64	72
9	9	18	27	36	45	54	63	72	81

Division

8 divided by 2 = 4   Another way of stating it $\frac{8}{2}$ = 4

The standard way of working the problem

$$\frac{4}{2\overline{)8}}$$

Problem-   573 divided by 5 or stated as 573 ÷ 5

Steps for Division   *with remainder*   *with decimal point*
1. Divide
2. Multiply
3. Subtract
4. Bring down
5. Start over with
   the new number.

```
        1 1 4                    1 1 4 . 6
    5 / 5 7 3                5 / 5 7 3 . 0
      - 5                      - 5
       0 7                      0 7
      -  5                      - 5
        2 3                      2 3
      - 2 0                      2 0
        3 remainder              3 0
                               - 3 0
                                   0
```

How to do the above problem

Step 1   5 / 5   5 goes into 5 one time
              5   1 times 5 is 5
              0   5 from 5 is 0

Step 2   5 / 5 7   bring the 7 down
              5   5 goes into 7 one time
             0 7   1 times 5 is 5
              5   5 from 7 is 2
              2

Step 3   5 / 5 7 3
              5
             0 7
              5
             2 3   bring the 3 down
             2 0   5 goes into 23 four times
              3   4 times 5 is 20
                           20 from 23 leaves a remainder of 3

A fraction expresses a part of a whole unit. Its counterpart is a decimal. One-half of a unit in fraction form is 1/2. In decimal form it is 0.5 .

Example

the line over a number means the number repeats itself forever

$$\frac{1}{4} = 0.25 \qquad \frac{2}{3} = 0.66\overline{66} \qquad \frac{5}{8} = 0.625 \qquad \frac{3}{4} = 0.75$$

In the fraction $\frac{5}{8}$ 5 is the numerator

8 is the denominator

(remember the one that is <u>d</u>own starts with "d ")

Different types of fractions

Proper fraction - the numerator is smaller than the denominator. $\frac{1}{4}$ and $\frac{7}{8}$

Improper fraction - the numerator is larger than the denominator. $\frac{7}{4}$ and $\frac{13}{8}$

Mixed number - contains a whole number and a fraction. $3\frac{1}{8}$ and $12\frac{5}{8}$

Improper fractions should be changed to whole or mixed numbers In other words the numerator should always be smaller than the denominator.

$\frac{7}{4}$ Divide the 7 by 4 = 1 with 3 left over

equaling $1\frac{3}{4}$

$3\frac{10}{5}$ Divide the 10 by 5 = 2 , then add the 2 to the 3 equaling 5

$12\frac{15}{8}$ Divide the 15 by 8 = 1 and 7 left over. Add the 1 to 12 equaling 13 with the 7 above the 8.

equaling $13\frac{7}{8}$

continued

Fractions

A fraction should be at its lowest terms. To do so the numerator and denominator must be divided by the same number.

$\frac{6}{8}$ both numbers can be divided by 2 = $\frac{3}{4}$

$\frac{10}{15}$ both numbers can be divided by 5 = $\frac{2}{3}$

$\frac{125}{200}$ both numbers can be divided by 25 = $\frac{5}{8}$

Addition of Fractions

Before adding or subtracting fractions, they have to have common denominators.

Example $\frac{1}{4} + \frac{3}{8}$

Step 1 Four goes into eight two times. So we will multiply both the numerator and denominator by 2. $\frac{1}{4}$ now becomes $\frac{2}{8}$

Step 2 Add the numbers 2 and 3 equaling 5. You do not add the denominators so bring down the 8.

$$\frac{2}{8}$$
$$+ \frac{3}{8}$$
$$\overline{\frac{5}{8}}$$

The following problem does not have a common denominator.

$$\frac{3}{8}$$
$$+ \frac{2}{5}$$
$$\overline{}$$

Step 1 Multiply the two denominators 8 X 5 = 40
This is the new denominator.
(continued on next page)

$\dfrac{3}{8}$ becomes $\dfrac{15}{40}$

Step 2 For the fraction 3/8 divide 8 into 40 = 5.
Multiply the 3 by 5 = 15.
The fraction 3/8 now becomes 15/40.

$\dfrac{2}{5}$ becomes $\dfrac{16}{40}$

Step 3 For the lower fraction 2/5 divide the 5
into 40 = 8. Multiply the 2 by 8 = 16.
The fraction 2/5 now becomes 16/40

$\dfrac{31}{40}$

Step 4 Add the two new numerators
15 and 16 = 31 . Now bring down the
denominator making the answer 31/40 .

A good way to set up this problem is

$$
\begin{array}{c|c}
\dfrac{3}{8} & \dfrac{15}{40} \\[2mm]
+\ \dfrac{2}{5} & \dfrac{16}{40} \\[2mm]
\hline
& \dfrac{31}{40}
\end{array}
$$

8 X 5 = 40
8 goes into 40 5 times
3 X 5 = 15

5 goes into 40 8 times
2 X 8 = 16

15 + 16 = 31

Lowest or Least Common Denominator (L C D)

If there are not many fractions to add or if the
fractions are not high numbers, multiplying them
together is the fastest way to get a common
denominator. If this is done with the fractions of

$$\frac{15}{16} + \frac{3}{5} + \frac{21}{64}$$

the common denominator would be 16 X 5 X 64 = 5120
This is a very high number to work with. To have a
lower common denominator do the following.

Step 1 Factor out each denominator

possible factors	1	2	3	4	5	6	7	8
16 =		2						8
5 =	1				5			
64 =		2		4				8

Step 2 For each denominator list out the factors used for the other denominators but not for the denominator in question. Then multiply them.

16 did not use the following factors that 5 and 64 did.
$$1 \times 4 \times 5 = 20$$
5 did not use the following factors that 16 and 64 did.
$$2 \times 4 \times 8 = 64$$
64 did not use the following factors that 16 and 5 did.
$$1 \times 5 = 5$$

Step 3 Multiply each member of each fraction by these new numbers 20, 64 and 5

$$\frac{15}{16} = \frac{15 \times 20}{16 \times 20} = \frac{300}{320}$$

$$\frac{3}{5} = \frac{3 \times 64}{5 \times 64} = \frac{192}{320}$$

$$\frac{21}{64} = \frac{21 \times 5}{64 \times 5} = \frac{105}{320}$$

Step 4 Now add the numerators of each fraction.
$$300 + 192 + 105 = 597$$

$$\frac{597}{320} = 1 \frac{277}{320}$$

Subtraction of Fractions

Subtraction requires the use of Lowest Common Denominators, just as addition does.

$$\frac{3}{4}$$
Subtract $\frac{1}{2}$

Change to lowest common denominator

$$\frac{3}{4}$$
$$\frac{2}{4}$$
_____
$$\frac{1}{4}$$

2 from 3 equals 1

Subtraction (continued)

$$\frac{3}{7} = \frac{15}{35}$$

$$\frac{2}{5} = \frac{14}{35}$$

$$\overline{\phantom{\frac{2}{5}} \quad \frac{1}{35}}$$

Step 1 Multiply the denominators together to get the common denominator. 7 X 5 = 35 the new denominator

Step 2 7 goes into 35 5 times
Multiply 3 times 3 = 15

Step 3 5 goes into 35 7 times
Multiply 2 times 7 = 14

Step 4 Subtract 14 from 15 = 1

Multiplication of Fractions

A common denominator is **not** needed for multiplying and dividing fractions. Usually when you multiply you expect to have a greater value than what you started with. Not true with fractions. Instead of saying "times" you should say "of". Example -- Three fourths of one half.

$$\frac{3}{4} \text{ of } \frac{1}{2} \qquad \text{or} \qquad \frac{1}{2} \text{ of } \frac{3}{4}$$

both mean the same

$$\frac{3}{4} \text{ X } \frac{1}{2} = \frac{3 \text{ X } 1}{4 \text{ X } 2} = \frac{3}{8} \qquad \frac{5}{16} \text{X} \frac{3}{4} = \frac{5 \text{ X } 3}{16 \text{ X } 4} = \frac{15}{64}$$

$$\frac{5}{8} \text{ X } 1\frac{3}{4} \left(\begin{array}{l} \text{Change whole and} \\ \text{mixed numbers to} \\ \text{improper fractions} \end{array} \right) \frac{5}{8} \text{ X} \frac{7}{4} = \frac{35}{32} = 1\frac{3}{32}$$

$$\frac{3}{4} \text{ X } 121 \frac{3}{4}$$

To change to an improper fraction multiply 121 X 4 = 484 and add 3 equaling 487.

$$\frac{3}{4} \text{X } \frac{487}{4} = \frac{1461}{16} = 91 \frac{5}{16}$$

16 goes into 1461
91 times with 5 left over

Fractions

Division of Fractions

Division is the inverse or opposite of multiplication. This can easily be done by inverting the divisor (the second fraction) and then multiplying straight across as we did in multiplication. To understand division of fractions state the following as, 1/2 goes into 5/8 one and one fourth times.

$$\frac{5}{8} \div \frac{1}{2} \qquad\qquad \frac{5}{8} \times \frac{2}{1} = \frac{10}{8} = 1\frac{1}{4}$$

Inverse the numerator
and denominator
then multiply

$$5\frac{13}{16} \div \frac{3}{4} \qquad 5 \times 16 = 80 \quad \text{then add 13 to 80} = 93$$

$$\frac{93}{16} \div \frac{3}{4} = \frac{93}{16} \times \frac{4}{3} = \frac{372}{48} = 7\frac{36}{48} = 7\frac{3}{4}$$

48 goes into 372 7 times
with 36 left over

Canceling to avoid large numbers - using the above problem

$$\overset{31}{\cancel{93}} \times \overset{1}{\cancel{4^1}} \qquad \begin{array}{cc} 31 & 3 \\ 4 & 4 \end{array} \qquad \begin{array}{l} 93 \text{ divided by } 3 = 31 \\ 16 \text{ divided by } 4 = 4 \end{array}$$

Changing Fractions to Decimals

To change a fraction to a decimal divide the numerator by the denominator.

$$\frac{1}{4} = \begin{array}{r} .25 \\ 4\overline{)1.00} \\ \underline{8} \\ 20 \\ \underline{20} \\ 0 \end{array} \qquad \frac{5}{8} = \begin{array}{r} .625 \\ 8\overline{)5.00} \\ \underline{48} \\ 20 \\ \underline{16} \\ 40 \\ \underline{40} \\ 0 \end{array}$$

Changing Decimals to Fractions

Some decimals can be changed to a fraction by dividing the decimal into the number one.

0.125   divide 1.000 by 0.125 = 8 = $\frac{1}{8}$ This only works if the result is a whole number.

If a number cannot be divided equaling a whole number, then a longer process is needed.

0.1875   Drop the decimal point and divide 1875 and 10000 by the same number. The larger the number you divide by, the fewer steps you will need.

$$0.1875 = \frac{1875}{10000} \quad \begin{array}{l} \frac{1875}{25} = 75 \to \frac{75}{25} = 3 \\ \\ \frac{10000}{25} = 400 \to \frac{400}{25} = 16 \end{array} \quad = \frac{3}{16}$$

$$0.375 = \frac{375}{1000} \quad \begin{array}{l} \frac{375}{25} = 15 \to \frac{15}{5} = 3 \\ \\ \frac{1000}{25} = 40 \to \frac{40}{5} = 8 \end{array} \quad = \frac{3}{8}$$

The following is best used for large decimal numbers. For illustration we will first do a small number.

0.625   Remove the decimal and $\frac{1,000}{625}$ = 1.6
        Divide into 1,000

Remove the decimal point from 1.6 and divide by a number that will result in a whole number. We will use the number 2.

$$\frac{16}{2} = 8 \text{ this is the denominator}$$

Now divide 8 into 1,000   Divide 125 into the original decimal

$$\frac{1,000}{8} = 125 \qquad \qquad \frac{625}{125} = 5 \text{ the numerator}$$

$\frac{5}{8}$ is the fraction that equals 0.625

Changing Decimals to Fractions

0.571428571   Remove the decimal and $\dfrac{1,000,000,000}{571,428,571} = 1.75$
               Divide into 1,000,000,000

Remove the decimal point and divide by a number that will result in a whole number

175 divided by 25 = 7   this is the denominator

Now divide 7 into 1,000,000,000

$$\dfrac{1,000,000,000}{7} = 142,857,143$$

Now divide this number into the original number

$\dfrac{571,428,571}{142,857,143} = 4$ this is the $\dfrac{4}{7} = 0.571428571$
                numerator

Another Example

0.640625 $\dfrac{1,000,000}{640625} = 1.56097561$ Remove the decimal point and round off to only 3 numbers.

there is more accuracy using a low number

Divide 156 by 2   Divide 78 into 1,000,000

$\dfrac{156}{2} = 78$ The new denominator $\dfrac{1,000,000,000}{78} = 12,820.51282$

Divide this new number into the original decimal number, with the decimal point removed.

$\dfrac{640,625}{12,812} = 50.0019513$ The new numerator $\dfrac{50}{78} = 0.640625$

$\dfrac{50}{78}$ can be reduced by dividing $\dfrac{25}{39}$
         50 and 78   by 2

To Round off to the nearest 64th of an inch.

Divide 64 by the denominator 39 = 1.64 . Then multiply 1.64 by the numerator 25 = 41

The fraction $\dfrac{25}{39}$ is rounded off to $\dfrac{41}{64}$

Fractions

Repeating Decimals from Fractions - Repeating numbers
sometimes results when a fraction is changed into a
decimal.

1/3 =0.3333333....3   5/6 = 0.833333...3
5/12 = 0.41666...6   26/33 = 0.787878...78

To show that a number repeats itself to infinity a line is drawn
over the number or numbers that repeat.

0.$\overline{3}$ 0.8$\overline{3}$ 0.41$\overline{6}$ 0.$\overline{78}$

There is a quick way to change a decimal that has repeating
numbers to a fraction.

Example- To change 0.$\overline{6}$ to a fraction do the following.

Extend to 4 decimal places = 0.6666 = N

N = 0.6666
10N = 6.666
10N - 1N = 6.666 - 0.6666 = 5.9994 Round off to 6
9N = 6

6 is now the numerator and 9 is the denominator

$\frac{9N}{9} = \frac{6}{9}$ divide the 6 and 9 by 3, equaling $\frac{2}{3}$

The equation form differs as to the repeating decimal

0.$\overline{3}$ 1 digit 1 place after the decimal point use 10N - N
0.8$\overline{3}$ 1 digit 2 places after the decimal point use 100N - 10N
0.41$\overline{6}$ 1digit 3 places after the decimal point use 1,000N -100N
0.$\overline{78}$ 2 digits 2 places after the decimal point use 100N - N
0.$\overline{037}$ 3 digits 3 places after the decimal point use 1,000N - N

(please refer to the following page for examples)

Repeating Decimals from Fractions (continued)

Example - $5/6 = 0.8\overline{3}$ use $100N - 10N$

$N = 0.8333$ $10N = 8.333$ $100N = 83.33$

$100N - 10N = 83.33 - 8.333 = 74.997$
round off to 75

$90N = 75$

75 in now the numerator and 90 is the denominator

$\dfrac{90N}{90} = \dfrac{75}{90}$ $N = \dfrac{75}{90}$ Divide both numbers by 15

$N = \dfrac{5}{6}$

Example - $5/12 = 0.41\overline{6}$ use $1000N - 100N$

$N = 0.4166$ $100N = 41.66$ $1,000N = 416.6$

$1,000N - 100N = 416.6 - 41.66 = 375$

$900N = 375$

$\dfrac{900N}{900} = \dfrac{375}{900}$ $N = \dfrac{375}{900}$ Divide both numbers
by 15

$N = \dfrac{25}{60}$ Divide both
by 5 $N = \dfrac{5}{12}$

Example - $26/33 = 0.\overline{78}$ use $100N - N$

$N = 0.7878$ $100N = 78.78$

$100N - N = 78.78 - .7878 = 77.992$
round off to 78

$99N = 78$

$\dfrac{99N}{99} = \dfrac{78}{99}$ Divide both
numbers by 3 $N = \dfrac{26}{33}$

see page 3:14 to change to 64th of an inch

Decimals are a continuation of the base 10 numbering system.

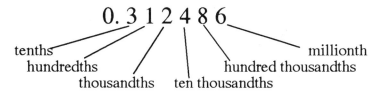

Examples: The decimal 0.31 would be stated as , "thirty one hundredths or thirty one percent ."
The decimal 0.435 would be stated as, " four hundred thirty five thousandths or forty three point five percent ."

Addition

```
  .5        3.3       .3875      143.6743
+ .5       +2.4      +.6274     + 96.8670
────       ────      ───────    ─────────
 1.0        5.7      1.0149      240.5413
```

Subtraction

```
  .8        1.52      3.12       341.6785
- .2       -.80      -2.81      - 256.4562
────       ─────     ─────      ──────────
 0.6        0.72      0.31        85.2223
```

Multiplication

Example

0.5 X 0.5 = 0.25   To better understand the multiplication of decimals the above should be stated as, "one-half of one-half" or" fifty percent of fifty percent."

2 spaces

```
  0.5
X 0.5
─────
  .25
```

count the numbers on the right side of the decimal point in the problem and move the decimal point that many times to the left in the answer.

2 spaces

Decimals

Example 0.25 X 3.5 = 0.875 is again easier to understand
if stated as," one forth of three and one-half."

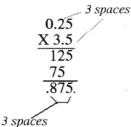

3 spaces

```
0.25
X 3.5
 125
  75
.875.
```

3 spaces

There are 3 numbers on the right
side of the decimal point in the
problem. After the numbers are
multiplied the standard way, the
decimal point is moved 3 spaces
to the left in the answer.

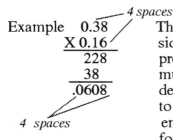

4 spaces

Example   0.38
          X 0.16
           228
            38
          .0608

4 spaces

There are 4 numbers on the right
side of the decimal point in the
problem. After the numbers are
multiplied the standard way, the
decimal point is moved 4 spaces
to the left. If there are not
enough numbers then insert a 0
for placement of the decimal.

Division

$$\frac{0.5}{0.5} = 1$$

The example is 0.5 divided by 0.5.
The best way to understand the
problem is to state it as, "one-half
goes into one-half, one time."

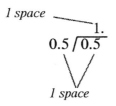

1 space

```
      1.
0.5 /0.5
```

1 space

Divide the standard way then move
the decimal point of the divisor all
way to the right . Count the
number of spaces moved and move
the decimal point the same number
of spaces to the right in the
dividend.

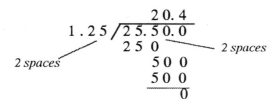

```
          2 0. 4
1 . 2 5 /2 5. 5 0. 0
          2 5 0          2 spaces
            5 0 0
            5 0 0
                0
```

2 spaces

Powers of 10 Scientific Notation

$0.0000000000001 = 1 \times 10^{-13}$ ten to the negative thirteenth power

or minus

$0.000000000001 = 1 \times 10^{-12}$ ten to the negative twelfth power

$0.0000000001 = 1 \times 10^{-10}$ ten to the negative tenth power

$0.0000001 = 1 \times 10^{-7}$ ten to the negative seventh power

$0.0001 = 1 \times 10^{-4}$ ten to the negative fourth power

$0.1 = 1 \times 10^{-1}$ ten to the negative first power

*(the following numbers are positive and
the positive sign (+) is never shown)*

$10 = 1 \times 10^{1}$ ten to the first power

$100 = 1 \times 10^{2}$ ten to the second power

$100000000 = 1 \times 10^{8}$ ten to the eighth power

$1000000000000 = 1 \times 10^{12}$ ten to the twelfth power

To change a decimal such as 0.0000587 to a whole number and a power of ten, count the number of places the decimal point is moved.

$$0.0\,0\,0\,0\,5\,.\,8\,7$$

$$5.87 \times 10^{-5}$$

To change a number such as 635,000,000.

$$6\,3\,5\,.\,0\,0\,0\,0\,0\,0\,.$$

$$635 \times 10^{6}$$

Even though the above number is correct, most numbers with an exponent will have a decimal.

$$6.35 \times 10^{8}$$

Exponents

Powers of 10 Scientific Notation

Exponents Power of 10 can be written 2 different ways.

$$8.35^{X\ 10^2} \qquad 8.35\ X10^{\,2} \qquad both = 835$$

The above is stated as 8.35 raised to the tenth power of two.

$$11.237\ X10^{\,-3} = 0.011237$$

The above is stated as 11.237 raised to the tenth power of negative three.

Multiplying Exponents Power of Ten

$$1.237\ X10^3\ X\ 1.65211\ X10^5 = 2.04366007\ X10^8$$

You add the exponents together when multiplying numbers with exponents.
If the tenth power was not used for the above example the answer would be 204,366,007

$$1.3\ X10^6\ X\ 1.237\ X10^5 = 1.6081\ X10^{11}$$

If the tenth power was not used the problem would be-

$$1,300,000\ X\ 123,700 = 160,810,000,000$$

$$1.67\ X10^{-6}\ X\ 6.347\ X10^{-7}\ 10.59949\ X10^{-13}$$

If the tenth power was not used the problem would be-

$$0.00000167\ X\ 0.0000006347 = 0.000000000001059949$$

Multiplying Exponents Power of Ten with Unlike Signs

$$1.375\ X10^5\ X\ 6.73\ X10^{-7} = 9.25375\ X10^{-2}$$

$$= 0.0925375$$

Exponents

Powers of 10   Scientific Notation

Multiplying Exponents Power of Ten with Unlike Signs

$$+2.561 \times 10^{-3} \times -3.31 \times 10^{5} = -8.47691 \times 10^{2}$$

Dividing Exponents Power of Ten

When dividing, subtract the lower exponent from the upper exponent.

$$\frac{25 \times 10^{5}}{12 \times 10^{2}} = 2.083333 \times 10^{3}$$

For the above problem use a calculator and do the following-

25 $\boxed{\text{EE key}}$ 5 $\boxed{\div}$ 12 $\boxed{\text{EE key}}$ 2 $\boxed{=}$ 2.083333×10^{3}

$$\frac{34 \times 10^{3}}{16 \times 10^{7}} = 2.125 \times 10^{-4}$$

In the above problem a positive 7 is subtracted from a positive 3 resulting with a negative 4.   *see page 4:10*

$$\frac{2.677 \times 10^{4}}{1.561 \times 10^{-6}} = 1.714926329 \times 10^{10}$$

4 subtract –6 = +10

For the above problem use a calculator and do the following-

2.677 $\boxed{\text{EE key}}$ 4 $\boxed{\div}$ 1.561 $\boxed{\text{EE key}}$ 6 $\boxed{+ - \text{key}}$ $\boxed{=}$
$$1.714926329 \times 10^{10}$$

Exponents

An exponent tells you how many times to multiply a number or expression times itself.

$$3^2 = 3 \times 3 = 9 \qquad 6^4 = 6 \times 6 \times 6 \times 6 = 1296$$

Using a calculator do the following 6 $\boxed{y^x \text{key}}$ 4 $\boxed{=}$ 1296

21^3 is stated as 21 to the cubed or 21 to the 3rd power

In the above example, 3 is the exponent of the base that is 21.

$$X^2 = X \cdot X \qquad\qquad 3^2 A^2 = 3 \cdot 3 \cdot A \cdot A = 9 A^2$$

Exponents in Fraction Form

$$\left(\frac{3}{4}\right)^3 = \frac{3 \cdot 3 \cdot 3}{4 \cdot 4 \cdot 4} = \frac{27}{64}$$ If you divide 27 by 64 it will equal 0.421875

The above can also be expressed as a decimal.

$$0.75^3 = .75 \times .75 \times .75 = .421875$$

$$0.07^3 = .07 \times .07 \times .07 = 0.000343 \text{ and as } 3.43 \times 10^{-4}$$

Multiplying Exponents

You can combine two powers as long as the base number is the same.

$6 + 4 = 10$

$$3^6 \times 3^4 = 3^{10}$$
$$(3 \times 3 \times 3 \times 3 \times 3 \times 3) \times (3 \times 3 \times 3 \times 3) = 3^{10}$$
$$729 \qquad \times \qquad 81 \qquad = 59,049$$

Using a calculator do the following-

3 $\boxed{y^x \text{ key}}$ 6 $\boxed{X}$ 3 $\boxed{y^x \text{ key}}$ 4 $\boxed{=}$ 59,049

Multiplying Exponents

$$6^{-2} = 0.027777778$$

$$6^{-2} \times 6^5 = 6^3 = 216 \qquad 6^5 = 7776$$

$$0.027777778 \times 7776 = 216$$

Using a calculator do the following-

6 $\boxed{y^x \text{ key}}$ 2 $\boxed{+ - \text{ key}}$ X 6 $\boxed{y^x \text{ key}}$ 5 $\boxed{=}$ 216

Dividing Exponents

You cancell out opposing exponents

$$\frac{3^4}{3^2} = \frac{3 \times 3 \times \cancel{3} \times \cancel{3}}{\cancel{3} \times \cancel{3}} = \frac{3 \times 3}{1} = 9 \qquad \frac{7^{12}}{7^4} = \frac{7^8}{1} = 5,764,801$$

$$\frac{8^3}{8^6} = \frac{\cancel{8} \times \cancel{8} \times \cancel{8}}{8 \times 8 \times \cancel{8} \times \cancel{8} \times \cancel{8}} = \frac{1}{8^3} = \frac{1}{512} \qquad \frac{9^{-12}}{9^7} = 9^{-19} = 7.402737006^{-19}$$

Using a calculator do the following-

8 $\boxed{y^x \text{ key}}$ 3 $\boxed{\div}$ 8 $\boxed{y^x \text{ key}}$ 6 $\boxed{=}$ 0.001953125 $= \frac{1}{512}$

Negative Exponents

Whole numbers

$$3^{-4} \text{ means} \qquad \frac{1}{3 \times 3 \times 3 \times 3} = \frac{1}{81} = 0.012345679$$

Using a calculator do the following-

3 $\boxed{y^x \text{ key}}$ 4 $\boxed{+ - \text{ key}}$ $\boxed{=}$ 0.012345679

Please note that when using a number with a value of more than 1 that has a negative exponent the results will be a number less than the value of one.

Negative Exponents

Less Than a Whole Number

$$0.06^{-3} = \left(\frac{6}{100}\right)^{-3} = \frac{1}{\left(\frac{6}{100}\right)^3} = \frac{1}{\frac{6 \times 6 \times 6}{100 \times 100 \times 100}}$$

$$\frac{\frac{1}{216}}{1,000,000} = \frac{1}{0.000216} = 4,629.62963$$

Using a calculator do the following-

.06 $\boxed{y^x \text{ key}}$ 3 $\boxed{+ -}$ = 4,629.62963

Radicals

Radicals

The radical sign $\sqrt{}$ means the root of a number. What number times itself, will equal this number? There are two variables, the index and the radicand. The index tells you how many factors or how many times a number should be multiplied times itself to equal this number, the radicand. If there is no index number, then it is assumed that the index is 2 and you are to square the number.

index $\diagup \sqrt[4]{21}$ — radicand

$\sqrt{9} = 3 \times 3 = 9$ $\sqrt[3]{27} = 3 \times 3 \times 3 \overset{21}{=} 27$

Most calculators have a $\boxed{\sqrt{X}}$ key and can do the square root of 9. However the key needed for finding the cube root or 4th root or higher is a $\boxed{\sqrt[x]{y}}$ key. With some calculators this will be the inverse or 2nd function of key $\boxed{y^x}$

Radicals (continued)

$\sqrt[3]{512}$ means that the cube root (3) is to be found for the radicand 512.

Using a calculator (Texas Instruments TI-30X) do the following-

512 [2nd key] [y^x key] 3 [=] 8

By using the 2nd function key what you have done is.

512 [$\sqrt[x]{y}$] 3 [=] 8   8X8X8 = 512

The inverse of the [y^x key] is the $\sqrt[x]{y}$ function.

$\sqrt[4]{81}$ = 3   3X3X3X3 = 81

$\sqrt[12]{576}$ = 1.6983133

Radicands with Exponents

$\sqrt[4]{9^3}$ = $\sqrt[4]{9X9X9}$ = $\sqrt[4]{729}$ = 5.196152423

Using a calculator do the following-

9 [y^x key] 3 [=] 729 [2nd key] [y^x key] 4 [=] 5.19615242

this is really the $\sqrt[x]{y}$ function

The above problem can also be done by dividing the index 4 by the exponent 3 = 1.3333
Now do the following with a calculator-

9 [$\sqrt[x]{y}$ key] 1.3333 [=] 5.196366503

If you have an index and exponent of the same value they cancel each other out to = 0.

Sets and Subsets

A $\subset$ B

A is a subset of B. Each element of A is an element of B and at least one element of B is not an element of A.
　　　Example 2,4,6 is a subset of
　　　　2,4,6,8,10,12,14
　　{2,4,6} $\subset$ {2,4,6,8,10,12,14}

A $\subseteq$ B

Each element in A is also an element of B. Also each element in B is an element in A. Thus
　　B $\subseteq$ A and A = B
Set A elements 1,2,3 equal Set B elements 1,2,3
A {1,2,3} $\subseteq$ B {1,2,3}

A $\cup$ B

A is a union with B. Elements of A or B or both, are all elements of the set.
Set A and B are a union and all their elements
　　　are of the same set 1,2,3,7,8,9
　　　A = 1,2,3 $\cup$ B = 7,8,9
　　　{1,2,3,7,8,9}

A $\cap$ B

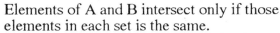

Elements of A and B intersect only if those elements in each set is the same.
　　The elements 5 and 6 of Set A intersect with
　　elements 5 and 6 of Set B.
　　{2,3,4,5,6} $\cap$ {5,6,7,8,9} = {5,6}

A $\subset \Omega$

A is a subset of a universal set.
　　Set A is a subset of the universal set of
　　all counting numbers.
　　Set A {1,2,3} is a subset of all counting
　　numbers

continued

Sets and Subsets

A ∩ Ω   Some of the elements in Set A intersect with a
               universal set.

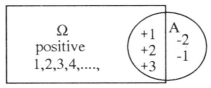

Set A has positive and negative numbers. The
Universal Set only has positive numbers.

A{-2,-1,+1,+2,+3} ∩   { +1,2,3,4,...,} = { 1,2,3,

A ∩ B = ∅   A intersects B = null set, { }

A ∈ of the Alphabet.   A is and element of the alphabet

Z ∈ A = Z is an element of the Set A

Z ∉ A = Z is not an element of Set A

{ } =   The set whose members are the set 1,2,3. { 1,2,3 }

∅ means the same as { } = an empty or null set

Sets and Subsets

Variable = Any letter or symbol that is used to denote any element of a specified set.

Finite Set = A set containing either no elements or a definite number of elements.

{ G,H,I,J }

Infinite Set = A set containing an unlimited number of elements, for example " all counting numbers"

A = { n | n = A equals the set n "such that or satisfying the condition of" n.

$\overline{A}$ = The complement of Set A

 = The area not A

Sets and Subsets

Venn Diagrams (circles)

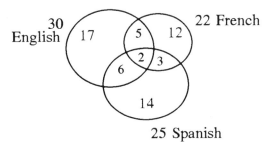

25 Spanish

Using the above Venn Diagram ;

There are 77 teachers; 30 speak English
22 speak French
25 speak Spanish
5 speak English and French
6 speak English and Spanish
3 speak French and Spanish
2 speak all 3 languages

To find how many speak only one language, for
example only English, add 5 + 2 + 6 = 13. Subtract
this from the total of 30 English speaking teachers and
you have 17 that only speak English.

You can also make the following statements.

- { 2 } ⊂ of all teaches
(The 2 teachers who speak all 3 languages are a Subset
of the Set of Teachers.)
- { 30 teachers} ∪ { 22 teachers } ∪ { 25 teachers} =
{ 77 teachers }
(The union of all teachers equal 77 teachers.)
- {English & Spanish}∩ { Spanish & English } = { 6 }
(The intersection of English and Spanish equals 6 who
speak English and Spanish.)
- 1 English Teacher ∈ { 30 English Teachers}
(One English Teacher is an element of the set of
English Teachers.)

Binary Number System

The computer age has made the very useful Binary Number System come to life. When a number is used we use a "1" and to show it is not used we use a "0". By starting with the value of 1 and doubling we can make the value of any number.

Base 10 system ➤ 128 64 32 16 8 4 2 1

Binary System
	128	64	32	16	8	4	2	1		
	0	0	0	0	0	1	0	0	=	4
	0	0	0	1	1	0	0	0	=	24
	0	1	1	0	0	0	0	1	=	97

Each position to the left of 1 doubles the preceding position.

to make the number 3
$$\frac{8\ 4\ 2\ 1}{0\ 0\ 1\ 1}\ (2+1=3)$$

to make the number 13
$$\frac{8\ 4\ 2\ 1}{1\ 1\ 0\ 1}\ (8+4+1=13)$$

to make the number 67
$$\frac{64\ 32\ 16\ 8\ 4\ 2\ 1}{1\ \ 0\ \ 0\ \ 0\ 0\ 1\ 1}$$
$$(64+2+1=67)$$

Addition   Rules 0 + 0 = 0
                           0 + 1 = 1
                           1 + 0 = 1
                           1 + 1 = Two 1s , so a 0 is put below and
                                   a 1 is carried to the
                                   next position to the left

		8	4	2	1
5	=	0^1	1^1	0^1	1
+ 3	=	0	0	1	1
8	=	1	0	0	0

1st step- there are two 1s so you carry over to the next column because 1 & 1 make 2.

2nd step

3rd step

4th step there is only one 1 so you carry it down to the answer row

continued

Addition (continued)

	64	32	16	8	4	2	1
92 =	1	0^1	1^1	1^1	1	0	0
24 =	0	0	1	1	0	0	0
4 =	0	0	0	0	1	0	0
125 =	1	1	1	1	0	0	0

Subtraction   The top number stays the same, however each element of the subtrahend is changed to the opposite. The 0s are changed to 1s and the 1s are changed to 0s. You then do the same as in addition. This answer is then changed by taking the last 1 on the left and adding it to the unit 1 column on the right.

$$
\begin{array}{ll}
 & 16\ 8\ 4\ 2\ 1 \\
23 = 1\ 0\ 1\ 1\ 1 \quad \text{stays the same} \quad \longrightarrow 1\ 0\ 1\ 1\ 1 \\
-17 = 1\ 0\ 0\ 0\ 1 \quad \text{change the 1s \& 0s} = 0\ 1\ 1\ 1\ 0 \\
\overline{6} \qquad\qquad\quad \text{to the opposite} \longrightarrow 1\ 0\ 0\ 1\ 0\ 1 \\
\qquad\qquad\qquad \text{then add.} \longrightarrow 1 \\
\end{array}
$$

$$\overline{0\ 0\ 1\ 1\ 0} = 6$$

Now shift the left 1 down to the right and then add as you would in addition.

Multiplication

Rules

$$0 \times 0 = 0$$
$$0 \times 1 = 0$$
$$1 \times 1 = 1$$
$$1 \times 0 = 0$$

$$12 \times 5 = 60$$

	8	4	2	1.
12 =	1	1	0	0
X 5 =	0	1	0	1
60				

Now use the same form as you do for regular multiplication but use the rules as shown above.

```
        1 1 0 0
   X    1 0 1
        1 1 0 0 ············· 0x1=0 0x1=0 1x1=1 1x1=1
      0 0 0 0   ············· 0x0=0 0x0=0 0x1=0 0x1=0
    1 1 0 0     ············· 1x0=0 1x0=0 1x1=1 1x1=1
    1 1 1 1 0 0  = 60
```

Dividing binary numbers is such a complicated procedure it will not be attempted in this text.

Roman Numerals

1	I	60	LX	1,500	MD
2	II	70	LXX	1,750	MDCCL
3	III	80	LXXX	1,900	MCM
4	IV	90	XC	1,910	MCMX
5	V	100	C	1,925	MCMXXV
6	VI	101	CI	1,950	MCML
7	VII	150	CL	1,960	MCMLX
8	VIII	200	CC	1,990	MCMXC
9	IX	300	CCC	2,000	MM
10	X	400	CD	2,500	MMD
11	XI	500	D	3,000	MMM
15	XV	600	DC	4,000	MM$\underline{\text{M}}$M
20	XX	700	DCC		or M$\overline{\text{V}}$
25	XXV	800	DCCC	5,000	$\overline{\text{V}}$
30	XXX	900	CM	10,000	$\overline{\text{X}}$
40	XL	1,000	M	50,000	$\overline{\text{L}}$
50	L	1,100	MC	100,000	$\overline{\text{C}}$
				500,000	$\overline{\text{D}}$
				1,000,000	$\overline{\text{M}}$

Examples -

23	XXIII
87	LXXXVII
135	CXXXV
1997	MCMXCVII
1,387	MCCCLXXXVII
2,837	MMDCCCXXXVII
3,111	MMMCXI
5,320	$\overline{\text{V}}$CCCXX
10,234	$\overline{\text{X}}$CCXXXIV
75,000	$\overline{\text{LXXV}}$
87,367	$\overline{\text{LXXXV}}$MMMCCCLXVII
2,500,000	$\overline{\text{MMD}}$
3,673,583	$\overline{\text{MMMVICLXX}}$MMMDLXXXIII
or	$\overline{\text{MMMCCCCCCLXX}}$MMMDLXXXIII

+	add, plus, positive sign
−	subtract, minus, negative sign
X	multiply
•	multiply
*	multiply
cd	c multiplied by d
÷	divide
a/b	a divided by b
$\frac{n}{y}$	n divided by y
±	plus or minus
:	ratio of
::	proportional to
∴	therefore
∵	because
=	equals
≠	not equal to
<	less than
>	more than
≯	not greater than
≮	not less than
=	identical to, equals
>>	much greater than
<<	much less than
≦	less than or equal to
≧	greater than or equal to

(continued on next page)

Math Symbols

$\equiv$	similar to
$\approx$	nearly equals
$\doteq$	approaches
∞	infinity
$\sqrt{}$	square root
$\sqrt[n]{y}$	n root of y
o	degrees
′	minutes, prime
′	feet
″	seconds
″	inches
n!	factorial
n	repeats to infinity
$\overline{\pi}$	pi 3.1416
a^n	a times itself n times
$\rightarrow$	approaches the limit
$\cap$	arch of a circle
$\underline{V}$	equiangular
Δ	increment
$\underline{\underline{\wedge}}$	corresponds to
$\perp$	perpendicular
$\parallel$	parallel
$\sum$	summation
$\int$	integral sign
$\underline{\underline{m}}$	measured by

Algebra

Signed Numbers

Addition

When the numbers have the same sign (+ or -)
add the two numbers using the same sign.

```
 + 8        - 7
 + 2        - 3
 +10       -10
```

When two numbers have different signs the
result is the difference of the two numbers. The
result has the sign from the greatest number

```
 + 8        - 5
 - 2        + 4
 + 6        - 1
```

Subtraction

When subtracting, change the sign of the
subtrahend then use the principles of addition in
algebra

```
From:     + 7    Change to:    + 7
Subtract: - 2                  + 2
                               + 9
```

```
From:     - 5    Change to:    - 5
Subtract: + 3                  - 3
                               - 8
```

Think of a thermometer scale when subtracting positive and
negative numbers. Also, change the order of the two numbers.
Instead of +7 take away -2 equals +9, think
from -2 to +7 you go 9 spaces in a positive direction. In
another example, -5 take away +3 equals -8, think instead,
from +3 to a -5 you go 8 spaces in a negative direction.

Algebra

Multiplication

If like signs, the product is positive; when unlike signs the product is negative.

$$\begin{array}{cccc} +6 & -4 & +6 & -3 \\ +3 & -5 & -3 & +3 \\ \hline +18 & +20 & -18 & -9 \end{array}$$

When multiplying it is better to think of the two numbers as two different elements, like money and time. In the following example- depositing money is positive and the past is positive; withdrawing and the future are both negative.

$$(+) \qquad X \qquad (+) \qquad = \qquad (+)$$
If I deposited $3 per day in the past 6 days, today I have $18 more.

$$(-) \qquad X \qquad (-) \qquad = \qquad (+)$$
If I withdraw $3 per day in the future 6 days, today I have $18 more.

$$(+) \qquad X \qquad (-) \qquad = \qquad (-)$$
If I deposit $3 per day in the future 6 days, today I have $18 less.

$$(-) \qquad X \qquad (+) \qquad = \qquad (-)$$
If I withdrew $3 per day in the past 6 days, today I have $18 less

Division

The same as in multiplication - if like signs, the product is positive; unlike, the sign is negative

$$\frac{+6}{+2} = +3 \qquad \frac{-10}{-5} = +2 \qquad \frac{-5}{+5} = -1 \qquad \frac{+50}{-10} = -5$$

When logically understanding division you should look at it as the inverse of multiplication.

Algebra

Order of Operations of Grouped Expressions
>( ) parentheses
>[ ] brackets
>{ } braces

Order of Operations - Signs
>Multiplication
>Division
>Addition
>Subtraction
>When using the above Order of Operations you must keep in mind, the first operation involves the least amount of elements.

Algebraic expressions
>monomial - one term
>binomial - two terms
>trinomial - three terms
>polynomial - two or more terms

Simplifying Expressions

Basic Rules -What you do on one side of the equal sign you have to do to the other side.
-If there are unknowns, separate them from the knowns to one side of the equal sign.

Example $3a = 9$ $\dfrac{\cancel{3}\,a}{\cancel{3}} = \dfrac{9}{3}$ $a = \dfrac{9}{3}$ $a = 3$

Example $4a - 5 = 115$ $4a - 5 + 5 = 115 + 5$

$4a = 120$ $\dfrac{4a}{4} = \dfrac{120}{4}$ $a = 30$

Example $10a - 3 = 19a - 39$ $10a - 3 + 3 = 19a - 39 + 3$

$10a = 19a - 36$ $10a - 19a = 19a - 19a - 36$

$\dfrac{-\cancel{9}\,a}{-\cancel{9}} = \dfrac{-36}{-9}$ $a = 4$

Algebra

Simplifying Expressions

Basic Rule - When using the Order of Operations, keep in mind that you first do operations that involve the least number of elements.

$$300 \times \left\{ \left[\frac{(1 + .07)^{(5+1)}}{.07} \right]^{-1} \right\} -1$$

add 1 + .07 = 1.07
add 5 + 1 = 6

$$300 \times \left\{ \frac{\left[(1.07)^6 \right]^{-1}}{.07} \right\} -1$$

multiply 1.07 times itself 6 times
calculate 1.07 $\boxed{Y^X}$ 6 $\boxed{=}$ 1.5007

$$300 \times \left[\frac{(1.5007 - 1)}{.07} \right] -1$$

$$300 \times \left(\frac{.5007}{.07} \right) -1$$

.5007 divided by .07 = 7.1526

$$300 \times (7.1526 - 1)$$

$$300 \times 6.1526 = 1845.78$$

Finding Missing Values

Example 1   Find the value of m

$8 - 6 - 3 m = - 6 + m - 2 m$
combine like terms
$+ 2 - 3 m = - 6 - m$
get like terms on either side of the = sign
    by adding or subtracting
$+ 2 - 3 m + 3 m = - 6 - m + 3 m$
combine like terms
$+ 2 = - 6 + 2 m$
again get like terms on either side of the = sign
    by adding or subtracting
$+ 2 + 6 = - 6 + 6 + 2 m$
again combine like terms
$+ 8 = 2 m$
divide both sides of the = sign by 2
$m = 4$

Example 2   Find the value of p

$3 (p - 3) + 4 (2 p + 3) = 3 (3 p) + 12$
simplify expressions
$3 p - 9 + 8 p + 12 = 9 p + 12$
combine like terms
$11 p + 3 = 9 p + 12$
get like terms on either side of the = sign
    by adding or subtracting
$11 p + 3 - 3 = 9 p + 12 - 3$
again combine terms
$11 p = 9 p + 9$
again get like terns on each side of the = sign
$11 p - 9 p = 9 p - 9 p + 9$
combine terms
$2 p = + 9$
divide both sides by 2
$p = 4.5$

Algebra

Finding Missing Values *(continued)*

Example 3

(when x = -3)

y = 8x + 7
y = (8 X -3) + 7
y = -24 + 7
y = -17

Example 4

(when a = -32)

7a - 8y = 32
(7 X -32) - 8y = 32
-224 -8y = 32
-224 + 224 - 8y = 32 + 224
-8y = 256
$$\frac{-8y}{-8} = \frac{256}{-8}$$
y = -32

Example 5

(when a = -3)

4a - 3y = -18
(4X-3) -3y = -18
-12 -3y = -18
-12 + 12 -3y = -18 + 12
-3y = -6
$$\frac{-3y}{-3} = \frac{-6}{-3}$$
y = 2

Example 6

(when a = -14)

3a + 7y = 70 + y
3a + 7y - 7y = 70 + y - 7y
3a = 70 - 6 y
3 • -14 = 70 - 6y
- 42 = 70 - 6y
- 42 - 70 = 70 - 70 - 6y
- 112 = -6y
$$\frac{- 112}{- 6} = \frac{- 6y}{- 6}$$
y = 18.667

Multiplying Algebraic Terms

5 a X 6 b can also be stated as (5 a) (6 b)
or (5 a)•(6 b) or 5•a•6 •b
(5 a) (6 b) = 30 a b

Example (7 a) (- 5 b) (2 c) = -70 a b c

Example (5 a) (2 a) (3 b) = 30 a a b or $30 a^2 b$

Example 2 a (7 a + 3) = $14 a^2 + 6 a$

Algebra

Monomials have only a single term. A term usually starts with a plus sign or negative sign and if there is no sign then the term is considered positive (+).

Example $5\,c$, $+2\,a^2 b^2$, $-cd$, $s^3 t^2$

Binomials have two terms and are also called polynomials.

Example $2\,a - b$, $4\,a + c^2$, $y + x$

Trinomials have 3 terms and are also called polynomials.

Example $x + y - 2$, $3\,a + b + c$, $3\,a^4 + b^3 + c^2$

Polynomials have two or more terms. Binomials and Trinomials are polynomials.

Adding Monomials

$$\begin{array}{rrr}
+\,5a & +\,6a & +\,12\,s^2 t^3 \\
(+)\ +\,3a & (+)\ -\,9a & (+)\ -\ \ 4\,s^2 t^3 \\
\hline
+\,8a & -3a & +\ \ 8\,s^2 t^3
\end{array}$$

Subtracting Monomials

$$\begin{array}{rrr}
+\,5a & +\,6a & +\,12\,s^2 t^3 \\
(-)\ +\,3a & (-)\ -\,9a & (-)\ -\ \ 4\,s^2 t^3 \\
\hline
+\,2a & +15a & +\ \ 16\,s^2 t^3
\end{array}$$

Multiplying Monomials

$$(5a)\,(3a) = 15a^2 \qquad (6a)\,(-9a) = -54a^2$$

$$(12\,s^2 t^2)\,(-4\,s^2 t^3) = -48\,s^4 t^5$$

Dividing Monomials

$$\frac{a^6}{a^2} = \frac{a\cdot a\cdot a\cdot a\cdot \cancel{a}\cdot \cancel{a}}{\cancel{a}\cdot \cancel{a}} = a^4 \qquad\qquad \frac{-18\,a^5}{3\,a^2} = -6\,a^3$$

$$\frac{+\,12\,a^2 b^3}{-\ 8\,a b^3 c} = \frac{12 \div 4 = 3}{8 \div 4 = 2} = \frac{3\cdot a\cdot \cancel{a}\cdot \cancel{b}\cdot \cancel{b}\cdot \cancel{b}}{2\cdot \cancel{a}\cdot \cancel{b}\cdot \cancel{b}\cdot \cancel{b}\cdot c} = \frac{3\,a}{2\,c}$$

Algebra

Adding Polynomials

$$\begin{array}{r} 3\,a + 7 \\ (+)\ \underline{4\,a + 6} \\ 7\,a + 13 \end{array} \qquad \begin{array}{r} 2\,a^2 - b + 3 \\ (+)\ \underline{5\,a^2 - 7\,b + 2} \\ 7\,a^2 - 8\,b + 5 \end{array} \qquad \begin{array}{r} 4\,a^3 + 7\,b - 3 \\ (+)\ \underline{-6\,a^3 - 3\,b + 4} \\ -2\,a^3 + 4\,b + 1 \end{array}$$

Subtracting Polynomials

$$\begin{array}{r} 4\,a + 7 \\ (-)\ \underline{4\,a + 5} \\ +2 \end{array} \qquad \begin{array}{r} 3\,a^3 - 5\,b^2 - 3 \\ (-)\ \underline{6\,a^3 + 2\,b^2 + 4} \\ -3\,a^3 - 7\,b^2 - 7 \end{array} \qquad \begin{array}{r} a^2 - 2 \\ (-)\ \underline{2\,a^2 + 2} \\ a^2 - 4 \end{array}$$

Multiplying Polynomials

One method of multiplying Polynomials is to use the **FOIL** System, First, Outer, Inner, Last It works best with binomials (2 terms).

Example $(x + 5) \cdot (x + 2)$

$(x + 5)\ X\ (x + 2)$
first
outer
inner
last

first = x^2
outer = $2\,x$
inner = $5\,x$
last = 10

$x^2 + 2\,x + 5\,x + 10$ *combine like terms*
$x^2 + 7\,x + 10$

Example $(3\,a^3 + 4) \cdot (7\,a^4 - 5)$

$(3\,a^3 + 4) \cdot (7\,a^4 - 5)$
first
outer
inner
last

first = $21\,a^7$
outer = $-15\,a^3$
inner = $28\,a^4$
last = -20

$21\,a^7 + 28\,a^4 - 15\,a^3 - 20$

Multiplying Polynomials *(continued)*

The best way to multiply trinomials is to discard the FOIL method and structure the problem as in normal multiplication. However you start on the left side.

Example $(3a + 5b - 2c) \times (4a - 6b)$

$$
\begin{array}{l}
3a + 5b - 2c \\
4a - 6b \\
\hline
12a^2 + 20ab - 8ac \\
\quad\quad -18ab \quad\quad\quad -30b^2 + 12bc \\
\hline
12a^2 + 2ab - 8ac - 30b^2 + 12bc
\end{array}
$$

$4a \cdot 3a = 12a^2$
$4a \cdot 5b = 20ab$
$4a \cdot -2c = -8ac$
$-6b \cdot 3a = -18ab$
$-6b \cdot 5b = -30b^2$
$-6b \cdot -2c = +12bc$

Example $(2a + b + 7) \times (4a + 3b + 3)$

$$
\begin{array}{l}
2a + b + 7 \\
4a + 3b + 3 \\
\hline
8a^2 + 4ab + 28a \\
\quad\quad + 6ab \quad\quad\quad + 3b^2 + 21b \\
\quad\quad\quad\quad + 6a \quad\quad\quad\quad\quad 3b + 21 \\
\hline
8a^2 + 10ab + 34a + 3b^2 + 24b + 21
\end{array}
$$

$2a \cdot 4a = 8a^2$
$b \cdot 4a = 4ab$
$7 \cdot 4a = 28a$
$2a \cdot 3b = 6ab$
$b \cdot 3b = 3b^2$
$7 \cdot 3 = 21b$
$2a \cdot 3 = 6a$
$b \cdot 3 = 3b$
$7 \cdot 3 = 21$

Example $(3a^2 - 9a + 5) \cdot (2a^3 + 4a^2 - 7)$

$$
\begin{array}{l}
3a^2 - 9a + 5 \\
2a^3 + 4a^2 - 7 \\
\hline
6a^5 - 18a^4 + 10a^3 \\
\quad\quad + 12a^4 - 36a^3 + 20a^2 \\
\quad\quad\quad\quad\quad\quad - 21a^2 + 63a - 35 \\
\hline
6a^5 - 6a^4 - 26a^3 - a^2 + 63a - 35
\end{array}
$$

$3a^2 \cdot 2a^3 = 6a^5$
$-9a \cdot 2a^3 = -18a^4$
$2a^3 \cdot +5 = +10a^3$
$3a^2 \cdot +4a^2 = +12a^4$
$-9a + 4a^2 = -36a^3$
$+5 + 4a^2 = +20a^2$
$3a^2 \cdot -7 = -21a^2$
$-9a \cdot -7 = +63a$

Dividing Polynomials

Examples - $\dfrac{4a - ab}{a} = \dfrac{\cancel{a}(4-b)}{\cancel{a}} = 4-b$

$$\dfrac{12x^2 + 30x^2 - 12x}{6x} = \dfrac{\overset{2x}{\cancel{12x^2}} + \overset{+5x}{\cancel{30x^2}} - \overset{-2}{\cancel{12x}}}{\cancel{6x}} = 7x - 2$$

$$\dfrac{-12a^3b + 30ab^2 - 9ab}{-3ab} = \dfrac{\overset{4a^2}{\cancel{-12a^3b}} + \overset{-10b}{\cancel{30ab^2}} \overset{+3}{-\cancel{9ab}}}{\cancel{-3ab}} = 4a^2 - 10b + 3$$

Factoring   To determine an equation or number that when multiplied times itself, produces the product in question.

Factors   When two or more numbers are multiplied by each other they are the factor of the product. The factors of 36 are; 18 and 2, 12 and 3, 9 and 4, 6 and 6   36 and 1, (*this is a trivial factor and is not commonly used*) However the **Prime Factors** are 3,3,2 and 2. Prime factors are numbers that cannot have other numbers, when multiplied together, equal the prime number. The prime factors of 9 and 4 are 3X3X2X2. The Prime factors of 6 and 6 are 3X2X3X2, which is the same as for 9 and 4. (Both equal 36) Usually when factoring you use the highest factor. For 36 you would use 6.

Factoring Monomials

$36a^2 = 6a$ which can also be stated as $(6a)^2$
or $(6a)(6a)$
$144 m^2n^6 = 12 mn^3$ or $(12mn^3)^2$

$$\dfrac{a^4}{16} = \left(\dfrac{a^2}{4}\right)^2 \qquad \dfrac{100}{a^8} = \left(\dfrac{10}{a^4}\right)^2$$

Algebra

Factoring Binomials *(two term polynomials)*

$$9a^2 - 4b^2$$

The key is to ask yourself, what number times itself equals 9 ? (answer 3). What number times itself equals 4 ? (answer 2) Put these two numbers in the equation.

$$(3 \quad 2) \text{ X } (3 \quad 2)$$

There has to be a negative number so give the second 2 a minus sign.

$$(3 \quad 2) \text{ X } (3 \quad -2)$$

We know the 3 has to have an " a", so the 2 must have a " b".

$$(3a + 2b) (3a - 2b) \text{ Check the answer using FOIL.}$$

first $3a \text{ X } 3a = 9a^2$

outer $3a \text{ X } -2b = -6ab$

inner $2b \text{ X } 3a = +6ab$ $9a^2 - 6ab + 6ab - 4b^2$

last $2b \text{ X } -2b = -4b^2$

$$9a^2 - 4b^2$$

$$a^2 - 1 \quad \text{the factor is} \quad (a + 1) (a - 1)$$

$$a^2 b^2 - 25 \quad \text{the factor is} \quad (ab + 5) (ab - 5)$$

Algebra

Factoring Trinomials *(three term polynomials)*

Example- Factor $10a + 20b - 15c$

What single number times three others will equal 10, 20, and 15? (answer 5 times 2, 4 and 3) 5 will be the start of the factor.

$5 \times \underline{2} = 10 \quad 5 \times \underline{4} = 20 \quad 5 \times -\underline{3} = -15$

Now put these numbers in the factor.

$5 (2 \quad 4 \quad 3)$

Next put in the letters and signs.

$5 (2a + 4b - 3c)$

Example- Factor $9x^2 - 6x + 21$

What single number times three others will equal 9, 6 and 21 ? (answer 3 times 3, 2 and 7) 3 will be the start of the factor.

$3 \times \underline{3} = 9 \quad 3 \times \underline{2} = 6 \quad 3 \times \underline{7} = 21$

Now put the numbers in the factor.

$3 (3 \quad 2 \quad 7)$

Now put in the letters and signs

$3 (3x^2 - 2x + 7)$

Factoring Trinomials *(three term polynomials)*

Example- Factor $a^2 + 14a + 49$

What single number times three others, will not work. The closest you can get is 7 X 2 +14 and 7 X 7 = 49. However a^2 has been left out.

Try starting with a and multiply it times each element.

$a X a = a^2$ $a X 7 = 7a$

$a (a + 7)$ Stop here, you know this will not work so try squaring the equation.

$(a + 7)^2$

Test by working the problem using FOIL.

$(a + 7) (a + 7)$ First $a \cdot a = a^2$
                            Outer $a \cdot 7 = 7a$
                            inner $7 \cdot a = 7a$
                            Last $7 \cdot 7 = 49$

Combine like terms
        a^2 + $14a$ + 49

Example- Factor $4a^2 - 4ab + b^2$

The factor of 4 is 2, so you know that will be an element. a and b are both squared so you know squaring will be an element and there is a negative also involved. b will also have to be multiplied by itself to equal b^2.

$(2a - b)^2$

Logic
Boolean Algebra

1 = on   0 = off

AND Logic

A and B are needed for an output

A	B	Output
0	0	0
0	1	0
1	0	0
1	1	1

OR Logic

A or B or Both are needed for an output.

A	B	Output
0	0	0
0	1	1
1	0	1
1	1	1

NAND Logic

A <u>and</u> B <u>not</u> needed for and output. This is the opposite of the AND Logic.

A	B	Output
0	0	1
0	1	1
1	0	1
1	1	0

NOR Logic

A <u>or</u> B <u>not</u> needed for an output. This is the opposite of the OR Logic.

A	B	Output
0	0	1
0	1	0
1	1	0
1	1	0

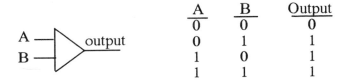

Perimeters

Four blocks all the same size produce different size perimeters.

$4 + 4 + 1 + 1 = 10$

$2 + 2 + 2 + 2 = 8$

Square

P = 4 X 1 side
P = 4 X 2 = 8

Rectangle

P = 2L + 2W

P = 2 X 6 + 2 X 2
P = 12 + 4
P = 16

Equilateral Triangle

P = 3 X Side

P = 3 X 3 = 9

Acute
or
Scalene
Triangle

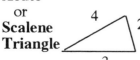

P = S+S+S

P = 4+3+2 = 9

Geometric Shapes

Circumference of a Circle   Pi=π = 3.141592654

C = Pi X Diameter

C = 3.1416 X 4 =12.5664
or
C = 2 X Pi X Radius

C = 2 X 3.1416 X 2

C = 12.5664

Diameter
4

Circumference of an Ellipse or Oval

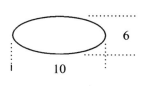

6

10

C = Pi X Average Diameter

$$C = 3.14 \text{ X } \frac{D1 + D2}{2}$$

$$C = 3.14 \text{ X } \frac{10 + 6}{2}$$

$$C = 3.14 \text{ X } \frac{16}{2}$$

A more accurate formula

C = 3.14 X 8
C = 25.12

$$C = Pi \sqrt{2(R_1^2 + R_2^2)}$$

$$C = 3.1416 \sqrt{2 \text{ X } (5^2 + 3^2)}$$

$$C = 3.1416 \sqrt{2 \text{ X } (5 \text{ X } 5 + 3 \text{ X } 3)}$$

$$C = 3.1416 \sqrt{2 \text{ X } (25 + 9)}$$

$$C = 3.1416 \sqrt{2 \text{ X } 34}$$

Please note that there are a variety of formulas for circumference and perimeter. These formulas have reasonable accuracy and are easy to work.

$$C = 3.1416 \sqrt{68}$$ *The square root of 68 is 8.2462*

C = 3.1416 X 8.2462 = 25.91 inches

Geometric Shapes

Area

Square
or
Rectangle

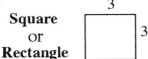

Side X Side
3 X 3 = 9 square inches

Triangle

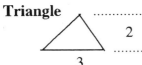

$$\frac{\text{Base X Height}}{2}$$

$$\frac{3 \text{ X } 2}{2} = 3 \text{ sq. inches}$$

see Triangle Section for more information

Parallelogram

Base X Height
6 X 3 = 18 square inches

Circle
(A flat surface)

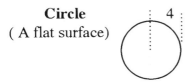

Pi X R^2
3.14 X (4 X 4)
3.14 X 16 = 50.24 sq. in.

see Circle Section for more information

Trapezoid

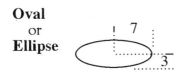

Average the top and base
then multiply the height

$$\frac{19 + 12}{2} \text{ X } 10$$

$$\frac{31}{2} \text{ X } 10 = 155 \text{ sq. inches}$$

Oval
or
Ellipse

This is not a 3 dimensional
shape, it is a flat surface

Pi X Average Radius

$$\text{Pi X } \left(\frac{R1 + R2}{2}\right)^2$$

$$3.14 \text{ X } \left(\frac{7 + 3}{2}\right)^2$$

$$3.14 \text{ X } \left(\frac{10}{2}\right)^2$$

$$3.14 \text{ X } 5^2$$

3.14 X 25 = 78.5 sq. in.

Geometric Shapes

Area (continued)

Polygons

Pentagon = 5 sides
Hexagon = 6 sides
Octagon = 8 sides
Nonagon = 9 sides
Decagon = 10 sides
Dodecagon = 12 sides

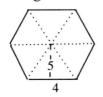

When each side of a polygon is equal, it is easy to figure the area.

Example - Hexagon
Make into 6 triangles then total together.

$$6 \times \frac{\text{Base} \times \text{Height}}{2}$$

$$6 \times \frac{4 \times 5}{2} = 6 \times \frac{20}{2}$$

$$6 \times 10 = 60 \text{ square inches}$$

Arc of a Circle

$(\text{Pi} \times R^2) \times \text{Percent of Circle}$
Determine the degrees by a protractor.

$$3.14 \times 5^2 \times \frac{70}{360}$$

$$3.14 \times 25 \times .1944 = 15.26 \text{ sq. inches}$$

Surface of a Pyramid

$$\frac{B \times H}{2} \times 4 \text{ sides}$$

$$\frac{4 \times 10}{2} \times 4$$

$$20 \times 4 = 80 \text{ square inches}$$

Surface of a Cone

Measure the slant height of the cone along its surface, this becomes **SH**

$$\frac{1}{2} \times (SH \times Pi \times R^2)$$

$$\frac{1}{2} \times (8 \times 3.14 \times (3 \times 3)$$

$$\frac{1}{2} \times (8 \times 3.14 \times 9)$$

$$\frac{1}{2} \times 226.08 = 113.04 \text{ square in.}$$

Area (continued)

Frustum Outer Surface

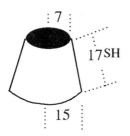

Average the Radius of the Base and Top-R1 and R2 Measure the height on the surface- this is SH

$$A = \left(\frac{R1 + R2}{2} \right) \times Pi \times SH$$

$$A = \left(\frac{15 + 7}{2} \right) \times 3.14 \times 17$$

$$A = \frac{22}{2} \times 3.14 \times 17$$

$$A = 11 \times 3.14 \times 17$$

$$A = 34.54 \times 17 = 587.18 \text{ sq. in}$$

Ball   (3 dimensional)

$$A = 4 \times Pi \, R^2$$

$$A = 4 \times 3.14 \times 5 \times 5$$

$$A = 12.56 \times 5 \times 5$$

$$A = 12.56 \times 25 = 314 \text{ sq. in.}$$

Sphere   (3 dimensional)
Ellipsoid

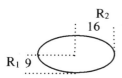

$$A = 4 \times Pi \times R_1 \times R_2$$

$$A = 4 \times 3.1416 \times 9 \times 16$$

$$A = 4 \times 3.1416 \times 144$$

$$A = 1809.56 \text{ square inches}$$

Geometric Shapes

Area

Trapezium

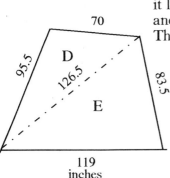

You can average the sides and then treat it like a rectangle. This will save time and the estimate will be 2-10% accurate. The closer each side is to being the same length the more accurate the estimate will be.

$$\frac{95.5 + 83.5}{2} = 89.5$$

$$\frac{70 + 119}{2} = 94.5$$

$$89.5 \times 94.5 = 8,457.5 \text{ square inches}$$

If all four sides and a measurement is across the area you treat it as if it was two oblique or obtuse triangles.

Triangle D   First Determine S $S = \dfrac{\text{Side 1} + \text{Side 2} + \text{Side 3}}{2}$

$$S = \frac{95.5 + 70 + 126.5}{2} = \frac{292}{2} = 146$$

$$\text{Area} = \sqrt{S\,[(\,S - \text{Side1})(\,S - \text{Side2})(S - \text{Side3})\,]}$$

$$\text{Area} = \sqrt{146\,[\,(146 - 95.5)(146 - 70)(146 - 126.5)\,]}$$

$$\text{Area} = \sqrt{146 \times 74,841}$$

$$\text{Area} = \sqrt{10,926,786} \qquad \textit{Calculate } 10,926,786 \sqrt{} \textit{ key} = 3,305.57$$

Area = 3,305.57 square inches for Triangle Area D

Triangle E   First Determine S $S = \dfrac{\text{Side 1} + \text{Side 2} + \text{Side 3}}{2}$

$$S = \frac{83.5 + 119 + 126.5}{2} = \frac{329}{2} = 164.5$$

continued on next page

Area Trapezium

Triangle E   (*continued*)

Area = $\sqrt{\text{S} \left[(\text{S} - \text{Side1})(\text{S} - \text{Side2})(\text{S} - \text{Side3}) \right]}$

Area = $\sqrt{164.5 \left[(164.5 - 83.5)(164.5 - 119)(164.5 - 126.5) \right]}$

Area = $\sqrt{164.5 \times 81 \times 45.5 \times 38}$

Area = $\sqrt{23,038,060.5}$

Area = 4,799.8 Square inches for Triangle Area E

Now add area D and Area E together

3,305.57 + 4799.8 = 8105.37 square inches

You can still find the area of a Trapezium, that you do not

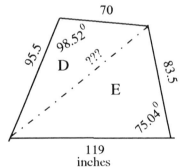

70

98.52°

95.5

D

???

E

83.5

75.04°

119
inches

have a measurement going across it, but do have a measurement of two opposing angles. Go to page 5 : 24 and work the formula as shown. This will then give you the length going across the trapezium. Then so back to the start of "Area for a Trapezium."

Geometric Shapes

Area

Rhombus

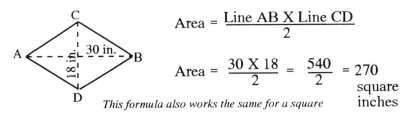

$$Area = \frac{Line\ AB \times Line\ CD}{2}$$

$$Area = \frac{30 \times 18}{2} = \frac{540}{2} = 270 \text{ square inches}$$

This formula also works the same for a square

Treat the rhombus shape as if it was two oblique triangles, if you only know the length of each side. See the section, "Finding Unknowns of an Oblique Triangle."

Circular Segment

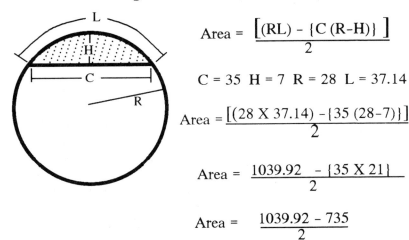

$$Area = \frac{\left[(RL) - \{C(R-H)\}\right]}{2}$$

$$C = 35 \quad H = 7 \quad R = 28 \quad L = 37.14$$

$$Area = \frac{[(28 \times 37.14) - \{35(28-7)\}]}{2}$$

$$Area = \frac{1039.92 - \{35 \times 21\}}{2}$$

$$Area = \frac{1039.92 - 735}{2}$$

$$Area = \frac{304.92}{2} = 152.46 \text{ square inches}$$

Geometric Shapes

Volume

Cube

$V =$ (side)^3

$V =$ 4 X 4 X 4 = 64 cubic in.

Rectangular Solid

$V =$ Length X Width X Height

$V =$ 5 X 3 X 6 = 90 cubic in.

Cylinder

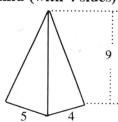

$V =$ Pi X R^2 X Height

$V =$ 3.14 X 4 X 4 X 12
3.14 X 16 X 12
$V =$ 3.14 X 192 = 602.88
cubic inches

Pyramid (with 4 sides)

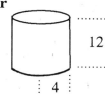

$V = \dfrac{1}{3}$ X Area of Base X Height

$\dfrac{1}{3}$ X (4 X 5) X 9

$\dfrac{1}{3}$ X 20 X 9

$\dfrac{1}{3}$ X 180 = 60 cubic inches

Cone

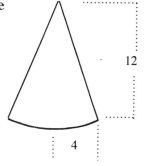

$\dfrac{1}{3}$ X(Pi X R^2 X Height)

$\dfrac{1}{3}$ X 3.14 X (4 X 4) X 12

$\dfrac{1}{3}$ X 3.14 X 16 X 12

$\dfrac{1}{3}$ X 3.14 X 192

$\dfrac{1}{3}$ X 602 88 = 200.96
cubic inches

Volume of Frustums

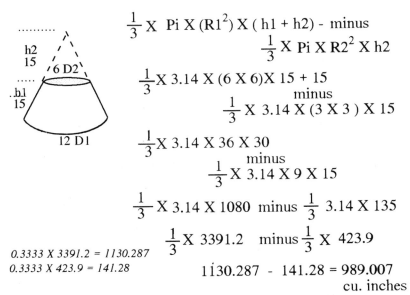

$$\frac{1}{3} \text{ X Pi X } (R1^2) \text{ X } (h1 + h2) - \text{minus}$$
$$\frac{1}{3} \text{ X Pi X } R2^2 \text{ X } h2$$

$$\frac{1}{3} \text{X } 3.14 \text{ X } (6 \text{ X } 6)\text{X } 15 + 15$$
$$\text{minus}$$
$$\frac{1}{3} \text{ X } 3.14 \text{ X } (3 \text{ X } 3) \text{ X } 15$$

$$\frac{1}{3}\text{X } 3.14 \text{ X } 36 \text{ X } 30$$
$$\text{minus}$$
$$\frac{1}{3} \text{ X } 3.14 \text{ X } 9 \text{ X } 15$$

$$\frac{1}{3} \text{ X } 3.14 \text{ X } 1080 \text{ minus } \frac{1}{3} \text{ } 3.14 \text{ X } 135$$

$$\frac{1}{3}\text{X } 3391.2 \quad \text{minus} \frac{1}{3} \text{ X } 423.9$$

0.3333 X 3391.2 = 1130.287
0.3333 X 423.9 = 141.28

$$1130.287 - 141.28 = 989.007$$
$$\text{cu. inches}$$

Height of Frustums

Note as in the above problem, one half the diameter of the base occurs at one half of the height. The height is directly proportional to the base.

Percent of Diameter occurs at 1 - (same percent) of height

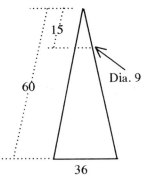

Example
.25 of 60 = 15.
.25 of the base 36 = 9

Volume of a Ball

$$\text{Volume} = \frac{4 \text{ Pi } R^3}{3}$$

$$V = \frac{4 \times 3.1416 \times 3.25 \times 3.25 \times 3.25}{3}$$

$$V = \frac{431.38095}{3} = 143.79365$$

Shortcut - You can find the volume of a ball by multiplying its diameter times itself 3 times then multiplying that by 0.5236
Using the above example- 6.5X6.5X6.5 = 274.625 X.5236 = 143.79365

Radius of a Ball with Volume Known

$$\text{Radius} = \sqrt[3]{\frac{V \times 3}{4 \text{ XPi}}} \quad = \quad \text{Radius} = \sqrt[3]{\frac{143.79 \times 3}{4 \times 3.1416}}$$

$$\text{Radius} = \sqrt[3]{\frac{431.37}{12.5664}} \quad = \quad \text{Radius} = \sqrt[3]{34.32725}$$

On your calculator do the following, enter 34.32725, then push the 2nd function key or the inverse key. Then push the Y^x key which is the $^x\sqrt{Y}$ function, followed by pushing the number 3. After pushing the = key, 3.2499 will appear on the screen.

$$\sqrt[3]{34.32725} = 3.2499$$

Volume of Spheres

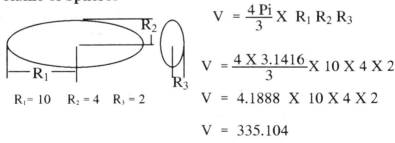

$$V = \frac{4 \text{ Pi}}{3} \times R_1 R_2 R_3$$

$$V = \frac{4 \times 3.1416}{3} \times 10 \times 4 \times 2$$

$$V = 4.1888 \times 10 \times 4 \times 2$$

$$V = 335.104$$

$R_1 = 10 \quad R_2 = 4 \quad R_3 = 2$

Please note that 4.1888 is a constant of the formula and does not change. The formula for volume of a ball and for a sphere are the same, they are just expressed differently. They both have the same results.

Geometric Shapes

Circle - Square Relationship

5

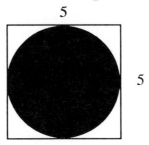

5

Area
-The area of the square is 5 X 5 = 25

-The area of any circle is 78.54 % of the square that it
     fits into. Therefor .7854 of 25 is 19.635
-When you multiply .7854 X 4 sides of a square you
     get 3.1416 (Pi).

-Pi X R^2= area of a circle
3.1416 X (2.5 X 2.5) = 3.1416 X 6.25 = 19.635
         or
.7854 X (Diameter^2)= area of a circle
.7854 X (5 X 5) = .7854 X 25 = 19.635

Circumference
C = Pi X Diameter   3.1416 X 5 = 15.708

or $\dfrac{Pi}{4}$ X Perimeter of Square = Circumference of the circle in the box

$\dfrac{3.1416}{4}$ X (5 X 4 sides) = 15.708

Volume of a Ball (also see previous page)
V = Circumference/6 X Diameter^3

$V = \dfrac{3.1416}{6}$ X 5 X 5 X 5

V = 0.5236 X 125 = 65.45 Cubic inches

Geometric Shapes

Circle - Finding Unknowns

To find the Radius with Area Known

$$R = \sqrt{\frac{\text{Area}}{\text{Pi}}}$$

Example- Find the radius of a circle with an area of 150.8 square inches?

$$R = \sqrt{\frac{150.8}{3.1416}} = \sqrt{48} = \text{The square root of 48 is 6.9283}$$

To find the Radius/Dia. with Circumference Known

$$R = \sqrt{\frac{\text{Circumference}}{3.1416}}$$

Example- Find the Diameter with a circle that has a circumference of 43.53 inches.

$$\text{Dia.} = \frac{43.53}{3.1416} = \underset{\text{Diameter}}{13.856} \quad \frac{13.856}{2} = \underset{\text{Radius}}{6.928}$$

Circle - Increasing the Area-

$$\text{Increase} = \sqrt{R^2} \text{ X the area larger or smaller that you want}$$

Example- A circle has a radius of 4 inches. What radius would make the area 200 % more.

200% = 2.0

$$\sqrt{4 \text{ X } 4 \text{ X } 2.0} = \sqrt{16 \text{ X } 2.0}$$

the square root of 32 is 5.6569

$$\sqrt{32} = 5.6569 \text{ Radius}$$

To check the answer
3.1416 X 5.6569 X 5.6569 = 100.53
3.1416 X 4 X 4 = 50.26 the original size

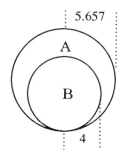

5.657

4

Circle A is twice the area of the smaller Circle B. However its diameter or radius is only 41 % bigger. In other words, if you wanted a pipe to carry twice as much water you would need a pipe 41 % bigger.

Geometric Shapes

Elements of a Circle

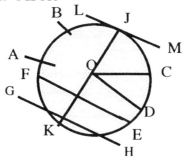

Arc - On the surface of the circumference, such as between A B .

Sector - A part of a circle shown as O C D .

Cord - A straight line cutting into a circle going to any two points, as shown as E F .

Secant - A line cutting across the circle, shown as GH.

Diameter - The distance across the circle shown as JK.

Tangent - A line touching a circle shown as LM.

Radius- A distance from the center to the edge shown as OC.

Parts of a Circle

A circle has 360 degrees (^{0})

Each degree has 60 minutes ($'$)
Each minute has 60 seconds ($''$)

A Grad is 1/ 400 of a circle and is 0.9 ^{0}

A Radian is 57.3 ^{0}

The statement - 26 degrees 4 minutes 30 seconds can be written as 26^0 4 $'$ 30 $''$

Geometric Shapes

The Three Angles

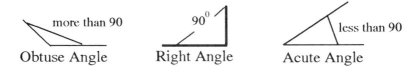

more than 90

Obtuse Angle

90^0

Right Angle

less than 90

Acute Angle

Angles

Add 23^0 32 ' to 95^0 51 '

$$95^0 \ 51 \ '$$
$$+ \ \underline{23^0 \ 32 \ '}$$
$$117^0 \ 83 \ '$$

there are 60 ' (minutes) to a degree so change
the answer to 118^0 23 '

Subtract 18^0 31 ' from 90^0
Change 90^0 to 89^0 60 '

$$- \ \underline{18^0 \ 31 \ '}$$
$$71^0 \ 29 \ '$$

Multiply 18^0 42 ' by 3

$$18^0 \ 42 \ '$$
$$\underline{X \qquad 3}$$
$$54^0 \quad 126 \ '$$ change to 56^0 6 '

Divide 33^0 21 ' into 3 parts

$$\frac{33^0 \ 21 \ '}{3} = 11^0 7 \ '$$

Geometric Shapes

Angles

Complementary Angles

To find the complementary angle of 42^0 18 ′ 13 ″

Change 90^0 to 89^0 59 ′ 60 ″

subtract 42^0 18 ′ 13 ″
$$\overline{ 47^0 \quad 41 ′ 47 ″}$$

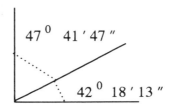

47^0 41 ′ 47 ″

42^0 18 ′ 13 ″

Supplementary Angles

To find the supplementary angle of 78^0 12 ′ 17″

Change 180^0 to 179^0 59 ′ 60 ″

subtract 79^0 12 ′ 17 ″
$$\overline{ 101^0 \ 47 ′ 43 ″}$$

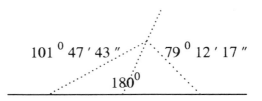

101^0 47 ′ 43 ″ 79^0 12 ′ 17 ″

180^0

Polygons - Finding the Length of Each Side

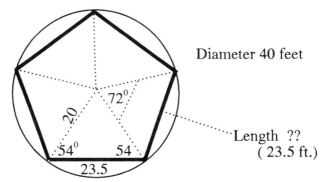

Diameter 40 feet

Length ??
(23.5 ft.)

Find how many degrees each portion takes up of a circle,
divide the number of sides into 360 degrees.

$$\frac{360}{5} = 72 \text{ degrees}$$

Find the value of the other two angles.

$$180 - 72 = 108 \qquad \frac{108}{2} = 54^0$$

If we know the diameter of the circle we know the length of
2 sides of each triangle.

$$\frac{\text{Diameter}}{2} = \frac{40}{2} = 20 \text{ feet}$$

To find the length of the remaining side -

$$\left(\begin{array}{c} \text{Number} \\ \text{of Sides} \end{array} \text{X} \begin{array}{c} \text{Diameter} \\ \text{of Circle} \end{array} \right) \text{X sine of} \left(\frac{360}{\begin{array}{c}\text{number}\\\text{of sides}\end{array}} \text{X} \frac{1}{2} \right)$$

$$(5 \text{ X } 40) \text{ X sine of} \left(\frac{360}{5} \text{ X} \frac{1}{2} \right)$$

$$(5 \text{ X } 40) \text{ X sine of} \left(72 \text{ X} \frac{1}{2} \right) = 200 \text{ X sine of } 36$$

on a calculator inter 36 then the [sine] key = .5877853 will appear

200 X .5877853 = 117.55   The last step divide this

$$\frac{117.55}{5} = 23.51 \text{ feet}$$

number by 5
(the number of sides)

Geometric Shapes

Triangles

Types according to sides
Equilateral - all three sides are equal
Isosceles - two sides are equal.
Scalene - none of the sides are equal

3 equal Sides   2 equal sides   no equal sides
Equilateral   Isosceles   Scalene

Types according to angles
Equiangular - all three angles are 60 degrees
Acute - each angle is less than 90 degrees
Right - one angle is 90 degrees
Obtuse - one angle is obtuse (greater than 90
degrees)
Oblique - a triangle with no right angle

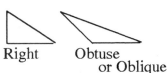

Equiangular   Acute   Right   Obtuse
or Oblique

Making a Right Angle

To make a right angle
layout a triangle main-
taining a 3 - 4 - 5 ratio
such as 6 - 8 - 10 or
15 - 20 - 25 , 12 - 16 - 20 etc.
Note that the two smaller
corresponding squares in the
example equals the area of the larger square.

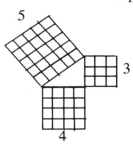

Angles of a Triangle Equal 180 Degrees.

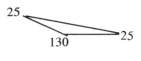

25 + 25 + 130 = 180   45 + 45 + 90 = 180   80 + 40 + 60 = 180

Triangles (continued)

Area of an Equilateral Triangle Using Length of Sides

Area = $\dfrac{S^2}{4}$ X $\sqrt{3}$

Area = $\dfrac{6 \text{ X } 6}{4}$ X $\sqrt{3}$

Area = $\dfrac{36}{4}$ X 1.732

Area = 9 X 1.732

Area = 15.589 square inches

Height of an Equilateral Triangle

Use the above triangle

Height = $\dfrac{S}{2}$ X $\sqrt{3}$

Height = $\dfrac{6}{2}$ X $\sqrt{3}$

Height = 3 X 1.732

Height = 5.196 inches

Equilateral Triangle Sizing

If you double the length of a side of an Equilateral Triangle, it is four times bigger. If you make a side 4 times longer, it is 16 times bigger.

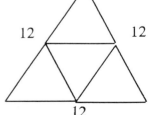

Geometric Shapes

Right Triangle

There are six functions
that are really ratios of
each other, that are used
to define the 3 angles and
3 sides that make up a **right**

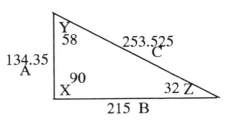

triangle. These six functions are sine, cosine, tangent, cosecant, secant, and cotangent. If we know any 2 elements of a right triangle we can determine the other 4 elements. (the right angle, 90 degrees, is not a variable and is not counted, but is assumed).

Angles not known but the 3 sides are known-

To fine the angle Y
    Divide side B by side C
      215 divided by 253.525 = .848042599
    Using a calculator inter the numbers
    .8481262, then the 2nd function key or
    the inverse key, then the sin key and
    57.999 will appear.

To find the angle Z
    Divide side A by side C
      134.35 divided by 253.525 = .5299280
    Using a calculator inter .5299280, then
    the 2nd or the inverse key, then the sin
    key and 32 will appear.

You have seen one of the functions of sine (sin). The following will give the formulas for all of the 6 functions however not until later will their use be discussed.

Sine   sin = the side opposite of the angle divided by the
                                         Hypotenuse
      sin Y = B divided by C
      sin Z = A divided by C
Using the above triangle
      sin Y = 215 divided by 253.525 = .848042599
      sin Z = 134.35 divided by 253.525 = .5299280
You can also do this on a calculator if you know the angle of Y
      Enter 58 then the sin key and .848048 appears.

Geometric Shapes

Right Triangle

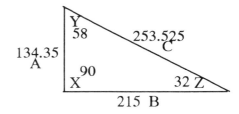

Cosine

cos = the adjacent side (part of the right angle) divided
by the Hypotenuse

cos Y = A divided by C
cos Z = B divided by C

Using the example triangle
cos Y = 134.35 divided by 253.525 = 0.529928015
cos Z = 215 divided by 253.5 = 0.848942599
Using a calculator, enter 58, the cos key and 0.529919 appears
Using a calculator, enter 32, the cos key and 0.848048 appears

Tangent

tan = the side opposite of the angle divided by the
adjacent side.   (the adjacent side is
tan Y = B divided by A   part of the right angle)
tan Z = A divided by B
Using the example triangle
tan Y = 215 divided by 134.35 = 1.60148976
tan Z = 134.35 divided by 215 = 0.6248837
Using a calculator, enter 58, the tan key and 1.600335 appears
Using a calculator, enter 32, the tan key and 0.6248694 appears

*The typical calculator does not have a function key for
cotangent, secant or cosecant.*

Cotangent

cot = the adjacent side divided by the side opposite of
the angle

cot Y = A divided by B
cot Z = B divided by A
Using the example triangle
cot Y = 134.33 divided by 215 = 0.624790698
cot Z = 215 divided by 134.33 = 1.600535993

Geometric Shapes

Right Triangle

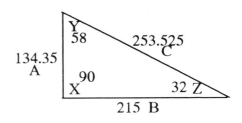

Secant

sec = the hypotenuse divided by the adjacent side

sec Y = C divided by A
sec Z = C divided by B

sec Y = 253.525 divided by 134.35 = 1.88704875
sec Z = 253.52 divided by 215 = 1.179186047

Cosecant

csc = the hypotenuse divided by the side opposite of
the angle

csc Y = C divided by B
csc Z = C divided by A

csc Y = 253.52 divided by 215 = 1.179186047
csc Z = 253.52 divided by 134.35 = 1.88704875

After studying the above you have probably noticed that each function is the same as another function.

    sine Y = cosine Z
    sine Z = cosine Y
    cosine Y = sine Z
    cosine Z = sine Y
    tangent Y = cotangent Z
    tangent Z = cotangent Y
    cotangent Y = tangent Z
    cotangent Z = tangent Y
    secant Y = cosecant Z
    secant Z = cosecant Y
    cosecant Y = secant Z
    cosecant Z = secant Y

Right Triangle

Finding Unknowns of a Right Triangle - Your triangle
should be drawn the same,
when comparing your
triangle to the example.

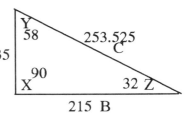

Unknown -
  Height Side A

Known -
  Both of the other 2 sides

$$A = \sqrt{C^2 - B^2}$$

$$A = \sqrt{(253.525 \times 253.525) - (215 \times 215)}$$

$$A = \sqrt{64,274.926 - 46,225}$$

$$A = \sqrt{18,049.926}$$

calculate 18,049.926 ⌐sq. root key⌐ *= 134.35*

$$A = 134.35$$

with some calculators do
18,049.926 ⌐inverse key⌐ ⌐Y^x key⌐ *2 = 134.35*

Unknown
  Height Side A

Known
  Side B and the
  smallest angle Z

$$A = B \times \tan Z$$

$$A = 215 \times \tan 32$$

Calculate 32 ⌐tan key⌐ *= .624869*

$$A = 215 \times .6248694$$

$$A = 134.35$$

Finding Unknowns of a Right Triangle ***(continued)***

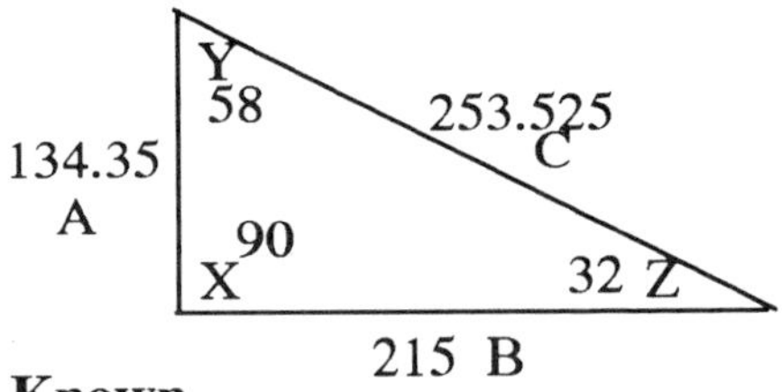

Unknown
Height Side A

Known
Side B and the largest angle Y

$$A = \frac{B}{\tan Y}$$

$$A = \frac{215}{\tan 58}$$

calculate 58 [tan key] *= 1.6003345*

$$A = \frac{215}{1.600335}$$

$$A = 134.3468$$

Unknown
Height Side A

Known
Side C and the smallest angle Z

A = C X sin Z

A = 253.525 X sin 32

calculate 32 [sin key] *= 0 .5299193*

A = 253.525 X 0.5299193

A = 134.3478

Unknown
Height Side A

Known
Side C and the largest angle Y

A = C X cos Y

A = C X cos 58

calculate 58 [cos key] *= 0 .5299193*

A = 253.525 X 0.5299193

A = 134.3478

**Finding Unknowns
of a Right Triangle**
(continued)

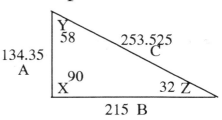

Unknown

Base Side B

$$B = C^2 - A^2$$

$$B = \sqrt{(253.525 \times 253.525) - (134.35 \times 134.35)}$$

$$B = \sqrt{64,274.9256 - 18,044.5489}$$

$$B = \sqrt{46,230.3767}$$

$$B = 215.01$$

Known

Both of the other 2 sides
C and A

calculate 46,230.3767 [sq. root key] *= 215.01*
see note below

Unknown

Base Side B

$$B = \frac{A}{\tan Z}$$

$$B = \frac{134.35}{\tan 32}$$

$$B = \frac{134.35}{.624869}$$

$$B = 215$$

Known

Side A and the smallest
angle Z

calculate 32 [tan key] *= .624869*

With some calculators you have to do the following in order to
do the square root function.

46230.3767 [2nd fnt key or inverse key] [Y^x key] 2 [=] 215

Geometric Shapes

**Finding Unknowns
of a Right Triangle**
(continued)

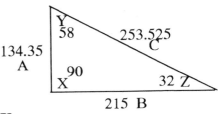

Unknown
Base Side B

Known
Height Side A and the
largest angle Y

$$B = A \times \tan Y$$

$$B = 134.35 \times \tan 58$$

calculate 58 $\boxed{tan\ key}$ = *1.600335*

$$B = 134.35 \times 1.600335$$

$$B = 215$$

Unknown
Base Side B

Known
Hypotenuse Side C and the
smallest angle Z

$$B = C \times \cos Z$$

$$B = 253.525 \times \cos 32$$

calculate 32 $\boxed{cos\ key}$ = *.848048*

$$B = 253.525 \times .848048$$

$$B = 215$$

Unknown
Base Side B

Known
Hypotenuse Side C and the
largest angle Y

$$B = C \times \sin Y$$

$$B = 253.525 \times \sin 58$$

calculate 58 $\boxed{sin\ key}$ = *.848048*

$$B = 253.525 \times .848048$$

$$B = 215$$

Geometric Shapes

Finding Unknowns
of a Right Triangle
(continued)

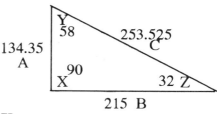

Unknown
  Hypotenuse Side C

Known
  Both of the other 2 sides
  A and B

$$C = \sqrt{A^2 + B^2}$$

$$C = \sqrt{(134.35 \times 134.35) + (215 \times 215)}$$

$$C = \sqrt{18{,}045.892 + 46{,}225}$$

$$C = \sqrt{64{,}270.892}$$

$$C = 253.517$$

calculate 64,269.549 $\boxed{sq.\ root\ key}$ = 253.517

Unknown
  Hypotenuse Side C

Known
  Height Side A and the
  smallest angle Z

$$C = \frac{A}{\sin Z}$$

$$C = \frac{134.35}{\sin 32}$$

calculate 32 $\boxed{sin\ key}$ = 0.529919

$$C = \frac{134.35}{.529919}$$

$$C = 253.529$$

Geometric Shapes

Finding Unknowns
of a Right Triangle
(continued)

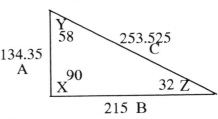

Unknown
    Hypotenuse Side C

$$C = \frac{A}{\cos Y}$$

$$C = \frac{134.35}{\cos 58}$$

$$C = \frac{134.35}{.529919}$$

$$C = 253.529$$

Known
    Height Side A and the
    largest angle Y

calculate 134.33 ÷ 58 $\boxed{cos\ key}$ *= 253.49*

Unknown
    Hypotenuse Side C

$$C = \frac{B}{\cos Z}$$

$$C = \frac{215}{\cos 32}$$

$$C = \frac{215}{.848048}$$

$$C = 253.523$$

Known
    Base Side B and the
    smallest angle Z

calculate 32 $\boxed{cos\ key}$ *= .848048*

Unknown
    Hypotenuse Side C

$$C = \frac{B}{\sin Y}$$

$$C = \frac{215}{\sin 58}$$

$$C = \frac{215}{.848048}$$

$$C = 253.523$$

Known
    Base Side B and the
    largest angle Y

calculate 58 $\boxed{sin\ key}$ *= .848048*

Geometric Shapes

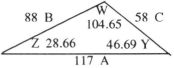

Finding Unknowns
of an Obtuse Triangle

Unknown	Known
Angle W	The length of all 3 sides

$$W = \text{Inverse } \cos \text{ of } = \frac{B^2 + C^2 - A^2}{2\,BC}$$

$$W = \text{Inverse } \cos \text{ of } = \frac{(88 \times 88) + (58 \times 58) - (117 \times 117)}{2 \times 88 \times 58}$$

$$W = \text{Inverse } \cos \text{ of } = \frac{7{,}744 + 3{,}364 - 13{,}689}{10{,}208}$$

$$W = \text{Inverse } \cos \text{ of } = \frac{11{,}108 - 13{,}689}{10{,}208}$$

$$W = \text{Inverse } \cos \text{ of } = \frac{-2{,}581}{10{,}208} = -0.2528409$$

Calculate .2528409 then $\boxed{+ - \text{key}}$ to make it a negative number

next the $\boxed{\text{inverse}}$ or $\boxed{\text{2nd function key}}$ then the $\boxed{\text{cos key}}$

equaling 104.64569

Using the above steps the following can also be figured.

Unknown	Known
Angle Y	The length of all 3 sides

$$Y = \text{Inverse } \cos \text{ of } = \frac{A^2 + C^2 - B^2}{2\,AC}$$

Unknown	Known
Angle Z	The length of all 3 sides

$$Z = \text{Inverse } \cos \text{ of } = \frac{A^2 + B^2 - C^2}{2\,AB}$$

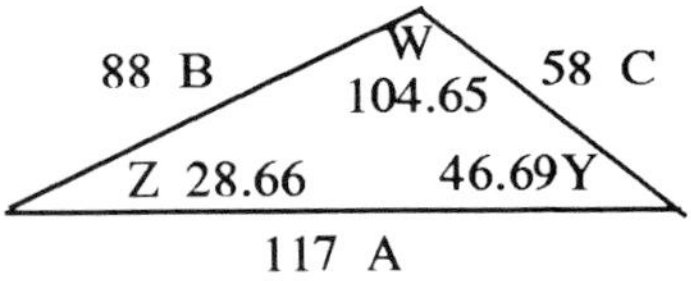

Finding Unknowns
of an Obtuse Triangle
(*continued*)

88 B   W   58 C
104.65
Z 28.66   46.69 Y
117 A

Unknown
     Side A

Known
     Sides B and C and Angle W

$$A = \sqrt{B^2 + C^2 - (2\,B\,C \times \cos W)}$$

$$A = \sqrt{(88 \times 88) + (58 \times 58) - (2 \times 88 \times 58 \times \cos 104.65)}$$

calculate 104.6 $\boxed{\cos\ \text{key}}$ = −.2529138

$$A = \sqrt{7{,}744 + 3{,}364 - (2 \times 88 \times 58 \times -.2529138)}$$

$$A = \sqrt{11{,}108 - (10{,}208 \times -.2529138)}$$

calculate 10,208 X .2529138 $\boxed{+\text{ - key}}$ = 2,581.7441

$$A = \sqrt{11{,}108 - (-2{,}581.7441)}$$

calculate 11,108 $\boxed{-}$ 2,581.7441 $\boxed{+\text{ - key}}$ = 13,689.744

$$A = \sqrt{13{,}689.744} \quad = \quad 117$$

Using the above steps the following can also be figured.

Unknown
     Side B

Known
     Sides A and C and Angle Y

$$B = \sqrt{A^2 + C^2 - (2\,A\,C \times \cos Y)}$$

Unknown
     Side B

Known
     Sides A and C and Angle Z

$$B = \sqrt{A^2 + C^2 - (2\,A\,C \times \cos Z)}$$

Geometric Shapes

**Finding Unknowns
of an <u>Obtuse</u> Triangle**
(*continued*)

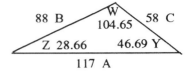

Unknown	Known
Side A	Side B and Angles W and Y

$$A = \frac{B \ X \ \sin W}{\sin Y}$$

$$A = \frac{88 \ X \sin 104.6}{\sin \ 46.69}$$ *Calculate 104.6* $\boxed{\sin key}$ *=.96748883*
Calculate 46.7 $\boxed{\sin key}$ *= .727653049*

$$A = \frac{88 \ X \ .96748883}{.727653049}$$

$$A = \frac{85.13901704}{.727653049} = 117$$

Using the above steps the following can also be figured.

Unknown   **Known**

Side A   Side C and Angles W and Z

$$A = \frac{C \ X \ \sin W}{\sin Z}$$

Side B   Side A and Angles W and Y

$$B = \frac{A \ X \ \sin Y}{\sin W}$$

Side B   Side C and Angles Y and Z

$$B = \frac{C \ X \ \sin Y}{\sin Z}$$

Side C   Side A and Angles W and Z

$$C = \frac{A \ X \ \sin Z}{\sin W}$$

Side C   Side B and Angles Y and Z

$$C = \frac{B \ X \ \sin Z}{\sin Y}$$

Geometric Shapes

Isosceles Triangle- can be treated as two right angles. By
doing so you can then find missing values using the
same formulas as you would a right triangle.

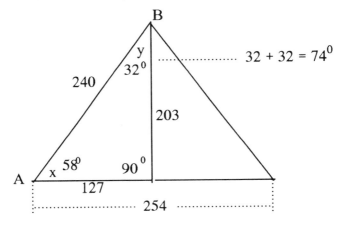

Mean - The average found by totaling all of the numbers then dividing them by the total number of items.

Mode - The number that most often appears in the group.

Median - The middle number; equal number of items appearing higher and lower then this number.

Range - The difference between the highest number in the group and the lowest number.

Mid-point - The average of the highest number and the lowest number.

Example

100		
90		
90	90 = Mode	
90		
85		
80	82 1/2 = Median	
65		
60		
50		
50		
———	$\frac{760}{10}$ = 76 Mean or Average	
760		

The range is 50

The midpoint of the range is 75

Percentile Rank - The percentage of people below the subject's score.

Example Mary scored 80 percent correct on her test. There were a total of 60 students, 30 scored lower; 10 scored the same as Mary and 20 scored higher.

$$\frac{\text{number that scored lower} + 1/2 \text{ of those scoring the same}}{\text{Total number of students}}$$

$$\frac{30 + (1/2 \text{ of } 10)}{60} = \frac{30 + 5}{60} = \frac{35}{60} = 58 \ 1/3 \text{ percentile rank}$$

Statistics and Probability

Geometric Mean

A method used to determine the probable position or value between two numbers, when each element is increasing more over each previous valuer.

Example - The population of a town is 50,000 in 1990 and in 2000 it is 81,444. This is a growth of 5% per year more than each previous year. To determine the population in 1995 you would use the geometric mean.

$$\sqrt{50,000 \text{ X } 81,444} \quad = \quad \sqrt{4,072,200,000}$$

Equaling 63,814 people in 1995

Please note that if you took the average of the two years it would be 65,722

$$\frac{50,000 + 81,444}{2} = 65,7222$$

The above can also be done by compounding the yearly increase. Keep in mind that each year the population increases 5 % over the previous year.

5 years - the year 1995
  1.05 X 1.05 X 1.05 X1.05 X 1.05 = 1.2762816

Using a calculator do the following
  1.05 Y^x key 5 = 1.2762816
  50,000 X 1.2762816 = 63,814 people in 1995

20 years - the year 2010

Using a calculator do the following
  1.05 Y^x key 20 = 2.6532977
  50,000 X 2.6532977 = 132,665 people in 1995

Weighted Average

When more importance is given to determined elements and all of the elements equal 100%.

Example - Instructor A grades in the following way.

Mid-term	25%
Final Test	50%
Class participation	15%
Homework	10%

If Mary scored the following; mid-term 80%, final exam 75%, class participation 80% and homework 75%. Her grade would be determined by the following.

Mid-term	.25 X 80 =	20
Final exam	.50 X 75 =	37.5
Class Participation	.15 X 80 =	12
Homework	.10 X 75 =	7.5
	Total Score =	77

Jeff takes the same class and his scores are; mid-term 95%, final exam 98%, class participation 30% and homework 50.%

Mid-term	.25 X 96 =	24
Final Test	.50 X 100 =	50
Class participation	.15 X 30 =	4.5
Homework	.10 X 50 =	5
	Total Score =	83.5

Jeff's total score is more than Mary's even though he seldom participated in class or did home work. His average grade (96+ 100+ 30 + 50 divided by 4) was 69. Mary's average grade was 77.5

Statistics and Probability

Standard Deviation - A measurement of variability from the mean (average) of a given group of numbers.

Example - Two different students test scores.

Bob	Bill
82	100
80	90
78	76
76	65
69	54
385	385

Step 1 Find the mean score

$$\frac{385}{5} = 77 \text{ mean} \qquad \frac{385}{5} = 77 \text{ mean}$$

Step 2 Determine the difference of each score from the mean.

	Bob			Bill	
Grades	Mean	Deviation	Grades	Mean	Deviation
82	77	+5	100	77	+23
80	77	+3	90	77	+13
78	77	+1	76	77	-1
76	77	-1	65	77	-12
69	77	-8	54	77	-23

Step 3 Square the deviations, then total the squared deviations

	Bob		Bill
Deviation	Deviation^2	Deviation	Deviation^2
+5	25	+23	529
+3	9	+13	169
+1	1	-1	1
-1	1	-12	144
-8	64	-23	529
	100		1372

Standard Deviation (continued)

Step 3 Find the average of the squared deviations by dividing the total by the number of deviations.

Bob	Bill

$$\frac{100}{5} = 20 \qquad\qquad \frac{1372}{5} = 274.4$$

Step 4 Find the square root of the averaged squared deviation.

$$\begin{array}{cc} 4.4721 & 16.566 \\ \sqrt{20} & \sqrt{274.4} \end{array}$$

The standard deviations are;
Bob = 4.4721   Bill = 16.566

Technical Formula for Standard Deviation

Example using Bill's grades in the previous problem.

$$\text{Standard deviation} = \sqrt{\frac{\Sigma (X - \mu)^2}{n}} = \sqrt{\frac{1372}{5}} = \sqrt{274.4} = 16.566$$

Σ = Sum Total
X = Each grade
μ = Mean grade
n = Number of Grades

Bill's Grades	Deviation from the Mean of 77		Deviation^2
X	$X - \mu$		$(X - \mu)^2$
100	100-77 =	+ 23	23 X 23= 529
90	90-77=	+ 13	13 X 13 = 169
76	76-77 =	- 1	-1 X -1 = 1
65	65-77 =	- 12	-12 X -12 = 144
54	54-77 =	- 23	-23X -23 = 529
Σ X=385			Σ $(X - \mu)^2$= 1372

Z Scores - The measurement of how many standard deviations
   a score or measurement is from the mean of a
   population or group.

Formula $Z_x = \dfrac{X - u}{\sigma}$

$$\dfrac{\text{the score being}}{\text{ranked}} = \dfrac{\text{raw score - mean}}{\text{standard deviation}}$$

Example - Taking the previous example of Bill's scores,
   we already know the standard deviation is
   16.567. we also know the mean is 77. His
   score of 90 ranks where as to a Z score?

$$Z = \dfrac{90 - 77}{16.57} = \dfrac{13}{16.57} = .78$$

Probability - This is another way of saying " making odds".
A favorable event is divided by the number of possible
outcomes. The answer is usually stated as favorable/
possible outcomes and can also be stated as a
percentage.

Formula Event =
$$\frac{\text{Favorable}}{\text{Total number of possible outcomes}}$$

Example - What are the chances a perfect coin will
turn up heads when tossed in the air ?

$$\frac{1}{2}$$ = 0.5 or stated as 1 out of 2
                                         or 1 to 2

Example - What are the chances a perfect coin will
turn up heads two times out of two tosses?

$$\frac{1}{2} \text{ X } \frac{1}{2} = \frac{1}{4} = 0.25 \text{ or stated 1 out of 4}$$

In the above example one event is independent
of another

Example - What are the chances of picking a red marble
out of a bag with 5 black ones and one red?

$$\frac{1}{6} = 0.1666 \text{ or stated as 1 out of 6}$$

Example - What are the chances of picking one red
marble out of a bag with 4 black ones and 2 red ?

$$\frac{2}{6} = \frac{1}{3} = 0.33 \quad \text{or stated 1 out of 3}$$

In the above example there were two favorable
out comes out of a possible 6.

Example - What is the probability of rolling two dice
at the same time with the same outcome, for
example each die shows a two?

$$\frac{1}{6} \text{ X} \frac{1}{6} = \frac{1}{36} \quad \text{stated as 1 chance in 36}$$

Statistics and Probability

Probability - Making Odds

Toss of a coin - There is a 50% chance of heads (win) and a 50% chance of tails (lose). This is a ratio of 1 to 2. If you were in a betting situation it would be said as, "one out of two."

Rolling dice - The chance of rolling a pair of dice and getting 2 is 1 chance in 36. When making odds you subtract the favorable outcome from the unfavorable 36 - 1 = 35. It would then be stated as "pays $35 to $1. In other words there are 35 chances of failure to a favorable chance of 1 in rolling a 2.

Horse Racing - The horse expected to win usually has odds of 2 to 1 and the horse most unlikely to win 20 to 1. However if the horse with 20 - 1 odds does win it pays $20 to every $1 you bet. To determine the horse most expected to win divide the last number by the first, resulting in a decimal. The higher the number the better the odds. However this is not the percent chance of the horse winning. To figure that, take the odds, total both numbers and divide that into the last number. For example the horse with 8 - 5 odds has 5/13 or 38.4% chance of winning.

Typical odds	Value of Odds	Percent chance of winning	
8 - 5	.655	5/13	38.4%
5 - 3	.60	3/8	37.5%
2 - 1	.50	1/3	33.3%
5 - 2	.40	2/7	28.5%
3 - 1	.333	1/4	25%
9 - 2	.222	2/11	18%
10 - 1	.10	1/11	9%
-	-		
-	-		
20 - 1	.05	1/21	4.7%

Probability

When understanding probabilities you also have to consider the number of possible combinations that can occur.

Example- A flip of a coin produces the following combination, heads (win) or tails (lose). There is a 1 to 2 or 50% chance of either occurrence, win or lose.

Example- When flipping a coin 2 times, what is the chance or probability of both being heads?

Both heads	w w	= .5 X .5	= .25	= 1/4
heads tails	w l	= .5 X .5	= .25	= 1/4
tails heads	l w	= .5 X .5	= .25	= 1/4
tails tails	l l	= .5 X .5	= .25	= 1/4

For more complex combinations it is best to setup the problem as follows.

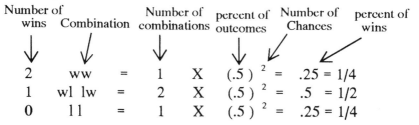

Number of wins	Combination	Number of combinations	percent of outcomes	Number of Chances	percent of wins
2	ww	= 1	X	$(.5)^2$ =	.25 = 1/4
1	wl lw	= 2	X	$(.5)^2$ =	.5 = 1/2
0	l l	= 1	X	$(.5)^2$ =	.25 = 1/4

The above table tells us the following-
  -there is a 25% chance of 2 heads out of 2 tosses.
  -there is a 50% chance of 1 head out of 2 tosses.
  -there is a 25% chance of no heads out of 2 tosses.

Example- A coin is flipped 3 times.

using a calculator enter
.5 $\boxed{Y^x \; key}$ 3 $\boxed{=}$ 0.125

3	www	= 1	X	$(.5)^3$ =	.125 1/8
2	wwl, wlw, lww	= 3	X	$(.5)^3$ =	.375 3/8
1	wll, lwl, llw	= 3	X	$(.5)^3$ =	.375 3/8
0	lll	= 1	X	$(.5)^3$ =	.125 1/8

Statistics and Probability

Probability and Combinations

Pascal's Triangle is very useful in determining probabilities and most important, the various number of combinations.

										Total possible combinations
				1						1
			1		1					2
		1		2		1				4
	1		3		3		1			8
1		4		6		4		1		16
1	5	10		10	5		1			32
1	6	15	20	15	6	1				64
1	7	21	35	35	21	7	1			128

Using Pascal's Triangle we can make the following chart.

N	0	1	2	3	4	5	6	7
0	1							
1	1	1						
2	1	2	1					
3	1	3	3	1				
4	1	4	6	4	1			
5	1	5	10	10	5	1		
6	1	6	15	20	15	6	1	
7	1	7	21	35	35	21	7	1

continues to infinity

Probability and Combinations (*continued*)

Using the chart from Pascal's Triangle we will now set up a problem of a coin tossed 5 times. To determine the combinations find 5 in the N column and go across finding 1,5,10,10,5 and 1 totaling 32 different combinations.

using a calculator enter
.5 $\boxed{Y^x \text{ key}}$ 5 $\boxed{=}$ *.03125*

No. of wins			
5	$1 \times (.5)^5$	= .03125	= 1/32
4	$5 \times (.5)^5$	= 0.15625	= 5/32
3	$10 \times (.5)^5$	= 0.3125	= 10/32
2	$10 \times (.5)^5$	= 0.3125	= 10/32
1	$5 \times (.5)^5$	= 0.15625	= 5/32
0	$1 \times (.5)^5$	= .03125	= 1/32

Please note that this step is really not necessary. By having the combinations listed 1,5,10,10,5,1 you can use them and the total of 32 to make the fraction form of each possible win. This works only with situations that are equal in the chance for a win or lose (50-50).

Example-

A wheel has three equal areas on it much like a pie chart and are colored red, blue, and green. If it is spun 5 times what are the chances of having 5 , 4, 3, 2, 1 or 0 wins? Wins, meaning the indicator falls on red.

This column should equal 100

5	$1 \times [(.333)^5]$	=	0.004095	= 0.41%
4	$5 \times [(.333)^4(.666)]$	=	0.409469	= 4.1%
3	$10 \times [(.333)^3(.666)^2]$	=	0.163788	= 16.4%
2	$10 \times [(.333)^2(.666)^3]$	=	0.327576	= 33%
1	$5 \times [(.333)(.666)^4]$	=	0.327576	= 33%
0	$1 \times (.666)^5$	=	0.131030	=13.1%

Keep in mind, the red (win) area is 33.3% of the wheel and the blue and green area (lose) is 66.6% of the wheel.

Probability and Combinations *(continued)*

The previous chart tells us there is less than a 1 % chance (0.41%) of having 5 red wins out of 5 attempts and a 13 % chance of not having any wins with 5 attempts.

Fibonacci Sequence

A progression of numbers with each new number being the sum of the previous two numbers.

$$1 \quad 1 \quad 2 \quad 3 \begin{array}{|ccc} 5 & 8 & 13 \\ 5 + 8 & = & 13 \end{array} \quad 21 \quad 34 \quad 55 \quad 89 \quad 144$$

Basically the next number is 162 % higher than the previous number.
It was first used in the 13th century by Leonardo Fibonacci. He made it to study reproduction and forever increasing number problems. Today it is seldom used because of modern calculators and computers.

The Language and Meaning of Statistics

You have to be careful on how a statistic is stated to get the real meaning. For example-
A sales company states that they had an increase in sales of $100,000 which was an increase over the previous year of 7% and they expect to increase sales the same amount over the next 5 years. Does this mean over the next 5 years sales will only increase $100,000. or will it increase $100,000 each year. If it increases 100,000 per year they would have the following growth

$$0.07 + 0.07 + 0.07 + 0.07 + 0.07 = .35 \text{ or } 35\%$$

If they had stated that they would have an increase of 7% over each previous year for the next 5 years, it would mean the following growth.

$$1.07 \times 1.07 \times 1.07 \times 1.07 \times 1.07 = 1.40 \text{ which}$$
would be 40% increase after 5 years.

Reliability- Single Possibilities of Success

When one item or function is in series with another the end result is failure of the series if any one of them fail.

Example - Suppose you are putting on an event with two different groups responsible for the out come. Group 1 has to work on various functions before Group 2 takes over and finishes the event. Group 1 tells you they are 90 % sure they will finish on time and Group 2 is 75 % sure. The probability or reliability of finishing on time can be figured by multiplying the probability of one group by the other group.

.90 X .75 = .675 = 67 1/2 % probability that the event will be done on time.

The more units or functions you put in series, one dependent on another for success, the less the reliability.

.90 X .75 = .675 = 67.5 % reliable
.90 X .75 X .85 = .574 = 57.4 % reliable
.90 X .75 X .85 X .99 = .568 = 56.8 % reliable

Reliability - Multiple Possibilities of Success.

When there is more than one function or unit that does not depend upon another for success the reliability increases.

Example - Mary lives in a metropolitan are. There are three highways that may or may not be jammed when she goes to work. She is informed that Highway A is clear 90% of the time, Highway B is clear 80% and Highway C is clear 70% of the time. What is the chance of Mary getting to work on time?

Multiple Possibilities of Success (continued)

- In the example one function does not depend on another for success. It only depends upon it, if it is not used.
- If Mary finds Highway A jammed she goes to Highway B, then if it is jammed she goes to Highway C.

Step 1 Figure the amount of time Highway A and B are used. Highway A is used 90% of the time, meaning 10% of the time Mary will try to use Highway B. Highway B is open 80% of the time, so multiply .10 X .80 equaling 0.08. Add to the 90% of the time Highway A is open and add the use of Highway B totaling 98%.

Step 2 Highway C is then used 2% of the time. Highway C is open 70% of the time. Multiply .02 X .70 = 0.014

Step 3 Add the .014 that Highway C will be used, to the use of Highway A and B of .98, totaling .994. Meaning 99.4% of the time Mary will get to work on time.

The above can be constructed into the following formula.

(A used) + (A jammed X B open) + (1.00 - Use of A and B) X (C open)

(.90) + (.10   X .80) + (1.00 - .98) X .70

.90 +   .08   + (.02 X .70)

.98   +   .014 = .994   =   99.4 % of the time Mary will get to work on time

Combinations and Permutations are very similar and are easily confused. To describe the difference we will look at an example. If there were a group of 5 people and you wanted to select 2 to go to a conference you could have 10 different couples to select from. Giving each member a letter to represent them. (AB), (AC), (AD), (AE), (BC), (BD), (BE), (CD), (CE), (DE). This is 10 different **combinations.** If you want to take pictures of 2 members at a time, from the group of 5, each member taking a turn setting on the right side then on the left side in each picture, you would have 20 different pictures. (AB), (BA), (AC), (CA), (AD), (DA), (AE), (EA), (BC), (CB), (BD), (DB), (BE), (EB), (CD), (DC), (CE), (EC), (DE), (ED). This is 20 different **permutations.**

Combinations - A subset of objects taken out of a given set of objects, each subset as selected cannot be rearranged or reused. When two or more elements appear together they can not appear together again. The order that they appear is of no importance but once they have appeared together they can not be counted again. For example the items A, B and C can be counted only once if there are to be three items. However if we only use two at a time there are three possible combinations AB, AC and BC. Please note that each item appears only once with the other elements.

Formula $$\frac{n\,!}{r\,!\ (n-r)\,!}$$

n = the number of items in the original set
r = the number of items in the subset
! = means to factor the number

examples of factoring
2 ! = 2 X 1 = 2
7 ! = 7X6X5X4X3X2X1 = 5040

Combinations (continued)

Example - Using the letters A B C D , how many
combinations can they make using 2 letters at a time?

$$C_2^{\,4} \;=\; \frac{4\,!}{2\,!\,(\,4-2\,)!} \;=\; \frac{4\times3\times2\times1}{(2\times1)\times(\,2!\,)} \;=\; \frac{24}{2\times1\times2\times1}$$

$$\frac{24}{4} = 6 \text{ combinations} \qquad \begin{array}{l} \text{AB,AC,AD,} \\ \text{BC,BD,DC} \end{array}$$

Example- An artist wants to make a drawing of 3 fruits
from a pile of 6 different fruits and he does not
want to reuse any combinations. How many can
can he make?

Formula $$\frac{n\,!}{r\,!\,(\,n-r\,)\,!}$$

$$C_3^{\,6} \;=\; \frac{6\,!}{3\,!\,(\,6\text{-}3)!} \;=\; \frac{6\times5\times4\times3\times2\times1}{(3\times2\times1)\times3!}$$

$$\frac{6\times5\times4\times3\times2\times1}{3\times2\times1\times3\times2\times1} = \frac{720}{36} = 20 \text{ Combinations}$$

This is an easier way to do it by cutting down the number of
times you have to multiply.

$$\frac{6\times5\times4\times\cancel{3}\times\cancel{2}\times\cancel{1}}{3\times2\times1\times\cancel{3}\times\cancel{2}\times\cancel{1}} = \frac{120}{6} = 20 \text{ Combinations}$$

Example - How many different combinations can be
made from the alphabet using 4 at a time?

cancel out duplicates

$$\frac{26!}{4!\,(\,26-4)!} = \frac{26!}{4!\,(22)!} \qquad \frac{26\times25\times24\times23\times\cancel{22\times21\times20\;\cdots\;\times1}}{4\times3\times2\times1\times\cancel{22\times21\;\cdots\;\times1}}$$

$$\frac{26\times25\times24\times23}{4\times3\times2\times1} \qquad \frac{358,800}{24} = 14,950$$

Permutation - Rearrangements of various items or elements, to get all possible arrangements from a predetermined group. The order of appearance of each item in the subset is important. Each item can reappear with another item as long as it is not in the same order. The items A, B and C can create 6 permutations using 3 at a time. (ABC), (ACB), (BCA), (BAC), (CAB), (CBA).

Formula $\quad P_r^n = \dfrac{n!}{(n-r)!}$

.

        n = number of items in the original set
        r = number of items in the subset
        ! = means to factorial Example
                      4 ! = 4X3X2X1 = 24

Example - Using the letters A,B,C and D , how many permutations can be made using only 2 at a time?

This permutation is stated as: n different things taken r at a time.

$P_2^4 = \dfrac{4!}{(4-2)!} = \dfrac{4X3X2X1}{2!} = \dfrac{24}{2X1} = \dfrac{24}{2}$ = 12 permutations

Another way of doing it:

P = 4 factorial 2 places (2 at a time)   P = 4 X 3 = 12

Example - Using the letters A,B,C,D,E and F how many permutations can be made using 3 at a time?

$P_3^6 = \dfrac{6!}{(6-3)!} = \dfrac{6X5X4X3X2X1}{3X2X1} = \dfrac{720}{6} = 120$

The quick way to do it.

P = 6 factorial 3 places (3 at a time) = 6X5X4 = 120

Permutations (continued)

Example - Using the letters A,B,C and D, how many
  permutations can be made using all of them at
  one time?

This permutation is stated as: n different things taken
n at a time.

Formula $P_4^{\ 4} =$ n! $P_4^{\ 4} =$ 4X3X2X1 = 24

Example - How many permutations/ arrangements can
  be made out of 8 pieces of fruit containing, 2
  bananas, 3 apples, 1 pear and 2 oranges

This permutation is stated as; n things taken p are alike,
  q are alike, r are alike, s are alike --

Formula

$$P = \frac{n!}{p!\ q!\ r!\ s!} \quad = \quad \frac{8!}{2!\ 3!\ 1!\ 2!}$$

$$\frac{8X7X6X5X4X3X2X1}{(\ 2X1)\ (3X2X1)\ (1)\ (2X1)} \ = \ \frac{40320}{24} = 1680 \text{ arrangements}$$

The easier way

$$\frac{8X7X6X5X4X\cancel{3}X\cancel{2}X\cancel{1}}{2X1\ X\ \cancel{3}X2X1\ X\ 1\ X\ \cancel{2}X\cancel{1}} \ = \ \frac{6720}{4} \ = 1680 \text{ arrangements}$$

Word Problems
Miscellaneous

Auditorium Size - The first row of seats in an auditorium has 10 seats. After the front row each row has 2 more seats than the row in front of it. How many seats are in the auditorium, if it has 30 rows?

$$(\text{number of rows})^2 + \left(\begin{array}{c}\text{seats in} \\ \text{first row}\end{array} \text{X} \begin{array}{c}\text{number} \\ \text{of rows}\end{array}\right) - (\text{number of rows})$$

$$(\ 30 \ )^2 + (10 \ \text{X} \ 30) - 30$$

$$(30 \ \text{X} \ 30 \ ) + 300 - 30 = 900 + 300 - 30 = 1170 \text{ seats}$$

This formula only works with 2 seats more per row. It will not work with any other configuration.

Determining Hourly Pay
   Mary made $ 1,425.00 in one week after working 51 hours. She worked 40 hours of regular time, 5 hours overtime at time and a half and 6 hours overtime at double time. How much does she normally make per-hour?

$$\begin{array}{c}(\text{Regular time}) \\ \text{hours}\end{array} + \begin{array}{c}(\text{Overtime X 1.5}) \\ \text{hours}\end{array} + \begin{array}{c}(\text{Overtime X 2}) \\ \text{hours}\end{array}$$

$$40 \ + \ 5 \ \text{X} \ 1.5 \ + \ 6 \ \text{X} \ 2$$

$$40 + 7.5 + 12 = 59.5$$

$$\frac{1425}{59.5} = \$ \ 23.95 \text{ per-hour}$$

Word Problems
Miscellaneous

A water pump truck holds a total of 500 gallons and is spraying
    20 gallons of water a minute. If it is half-full and
    continues to spray water at 20 gallons a minute, how
    many gallons will it take to fill the tank in 40 minutes?

Determine how many gallons will be sprayed while the
    tank is being filled.

$$40 \text{ minutes X 20 Gal. per minute} = 800 \text{ gallons}$$
$$\text{The tank is half-full } = \underline{250} \text{ gallons}$$
$$\text{Total } = \overline{1050} \text{ gallons}$$

How long would it take to fill the above tank if you fill it at 30
    gallons per minute?

$$\text{From the input of 30 gallons per minute}$$
$$\text{Subtract the amount going out } \underline{20 \text{ gallons}} \text{ per minute}$$
$$10 \text{ gallons}$$

Divide 250 gallons by 10 = 25 minutes

How many gallons per minute would you have to pump into
    the tank if you had to fill it in 10 minutes while it was
    still spraying water at 20 gallons a minute?

Determine how many gallons it will use while it is
    being filled.

$$10 \text{ minutes X 20 gallons per minute} = 200 \text{ gallons}$$
$$\text{then add the half empty tank } = \underline{250} \text{ gallons}$$
$$\text{total } = \overline{450} \text{ gallons}$$

Now divide the 450 gallons by 10 minutes = 45 gallons
per minute of water to fill the tank.

Miscellaneous Numbers

A swimming pool holds 16,830 gallons of water. Water out of
Pipe A flows at 10.6 gallons per-minute and out of
Pipe B it flows at 8.5 gallons per-minute. How long
will it take to fill the pool when both pipes are used
at the same time?

Pipe A = 10.6
Pipe B = 8.5
19.1 Gallons per minute.

To get number of hours, divide the number of gallons
in the pool by 19.1

$$\frac{16,830}{19.1} = 881.15 \text{ minutes} \qquad \frac{881.15}{60} = 14.69 \text{ hours}$$

A swimming pool has 3 pipes that can fill it. Pipe A will fill it
in 10 hours; Pipe B in 15 hours and Pipe C in 12 hours.
How long will it take if all 3 pipes are used?

Pipe A has the shortest time so base the Pipes B and C
on Pipe A.

Time = A + B + C

$$A = \frac{10}{10} \qquad B = \frac{10}{15} \qquad C = \frac{10}{12}$$

$$\frac{10}{10} = 1 \qquad \frac{10}{15} = 0.6667 \qquad \frac{10}{12} = 0.8333$$

Time = 1 + 0.6667 + 0.8333

Time = 2.5 hours to fill the pool using all 3 pipes.

Word Problems
Miscellaneous Numbers

One-half of a number is the same as two and one-half times the
same number minus forty. What is the number?

$$\frac{1}{2} N = 2\frac{1}{2} N - 40$$

get the unknown N on one side of the = sign

$$\frac{1}{2}N - 2\frac{1}{2} N = 2\frac{1}{2}N - 2\frac{1}{2}N - 40$$

$$\frac{1}{2}N - 2\frac{1}{2} N = 2\frac{1}{2}\cancel{N} - 2\frac{1}{2}N - 40$$

combine like terms

$$-2N = -40$$

$$\frac{\cancel{-2}N}{\cancel{-2}N} = \frac{-40}{-2}$$ A negative number divided by a negative
number results in a positive number

$$N = +20$$

Check

$$\frac{1}{2} \text{ of } 20 = 2.5 \text{ of } 20 - 40$$

$$10 = 50 - 40$$
$$10 = 10$$

The cost of maintenance on equipment is $ 6,280.00 and is
1.2 % of total sales. What is the total amount of sales?

Divide cost of maintenance by its percent of sales.

$$\frac{6,280}{0.012} = \$ 523,333.33 \text{ total sales}$$

Miscellaneous Numbers

Two thirds of a number is the same as 3 less than three fourths of it. What is the number?

$$\frac{2}{3} \text{ of N} = \frac{3}{4}\text{N} - 3$$

We could do this problem using fractions, as in the previous example, however the use of calculators makes it easier if we change the problem to decimals.

0.6667 N = 0.75 N - 3

Put the unknown N on one side of the equation

0.6667 N - **0.75 N** = 0.75 N - **0.75 N** - 3

0.6667 – 0.75 = – 0.08333

- 0.0833 N = - 3

$$\frac{-\cancel{0.0833}\,\text{N}}{-\cancel{0.0833}} = \frac{-3}{-0.0833}$$

A negative number divided by a negative number results in a positive number
-3 divided by - 0.0833 = 36

N = 36

Check

$$\frac{2}{3} \text{ of } 36 = 24 \quad \left(\frac{3}{4} \text{ of } 36\right) - 3 = 24$$

Word Problems
Miscellaneous Numbers

A is 4 more than B. C is 2 less than A. A + B + C = 18. What are the individual values of A,B, and C?

$$A = A \quad B = A - 4 \quad C = A - 2$$

$$A + (A - 4) + (A - 2) = 18$$

$$3 A - 6 = 18$$

$$3 A - 6 + 6 = 18 + 6$$

$$3 A = 24$$

$$\frac{\cancel{3} A}{\cancel{3}} = \frac{24}{3} = 8$$

$$A = 8$$
$$B = 8 - 4 = 4$$
$$C = 8 - 2 = 6$$

A recycle center collected 8,200 pounds of newspaper and cardboard. They collected 4 times more newspaper than cardboard. How many pounds of each did they collect?

$$4X = \text{Newspaper} \quad X = \text{Cardboard}$$

$$4X + X = 8,200$$

$$5X = 8,200$$

$$\frac{\cancel{5}X}{\cancel{5}} = \frac{8,200}{5} = 1640$$

$$X = 1,640 \text{ pounds of cardboard}$$
$$4X = 6,560 \text{ pounds of newspaper}$$

Word Problems

Miscellaneous Numbers

There are 27 adults in a room. There are twice as many women
as there are men. How many women and men are in
the room?

$$\text{men} \quad \text{women}$$
$$X + 2X = 27 \text{ people}$$

$$3X = 27$$

$$\frac{\cancel{3}X}{\cancel{3}} = \frac{27}{3} = 9$$

$$X = 9 \text{ men}$$
$$2X = 18 \text{ women}$$

Bob is 5 years older then Mary. The sum of their ages is 39.
How old is Bob and how old is Mary?

$$\text{Bob} \quad \text{Mary}$$
$$X + X - 5 = 39$$

$$2X - 5 = 39$$

$$2X - 5 + 5 = 39 + 5$$

$$2X = 44$$

$$\frac{\cancel{2}X}{\cancel{2}} = \frac{44}{2}$$

$$X = 22$$

Bob is 22 and Mary is 17

Check   22 + 17 = 39

Word Problems
Miscellaneous Numbers

A theater sold 380 tickets. The total proceeds from the tickets
 was $ 1,572.75. The student tickets sold for $ 3.25 and
 the adult tickets were $ 5.00. How many student and
 adult tickets were sold?

$$X = \text{adults} \quad 380 - X = \text{students}$$

$$5.00\,X + 3.25\,(\,380 - X\,) = \$\,1,572.75$$

$3.25\ X\ 380 = 1235.00$

$$5\,X + 1,235 - 3.25\,X = 1,572.75$$

$$5\,X + 1,235 - 1,235 - 3.25\,X = 1,572.75 - 1,235$$

$$5\,X - 3.25\,X = 337.75$$

$$1.75\,X = 337.75$$

$$\frac{\cancel{1.75}\,X}{\cancel{1.75}} = \frac{337.75}{1.75}$$ 337.75 divided by $1.75 = 193$

$$X = 193 \text{ adult tickets} \quad 380 - 193 = 187 \text{ student tickets}$$

The same previous problem from the above, however this time
 X will equal student tickets sold and $380 - X$ will equal
 adult tickets sold. (This gives us the opportunity to
 gain experience working with negative numbers)

$$3.25\,X + 5.00\,(\,380 - X\,) = \$\,1,572.75$$

$5\ X\ 380 = 1,900$

$$3.25\,X + 1,900 - 5\,X = 1,572.75$$

$$3.25\,X + 1,900 - 1,900 - 5\,X = 1,572.75 - 1,900$$

$$3.25\,X - 5\,X = -327.25 \qquad -1.75\,X = -327.25$$

$$\frac{-\cancel{1.75}\,X}{-\cancel{1.75}} = \frac{-327.25}{-1.75}$$ *A negative divided by a negative*
number results in a positive number

$$X = 187 \text{ student tickets}$$
$$380 - 187 = 193 \text{ adult tickets}$$

Miscellaneous Numbers

John and Mary are going to put money into a project that cost
$ 180,000.00. Mary is going to put in one-fourth of
whatever John puts in. (Mary will put in 20% and
John will put in 80%.) How much will each put in?

X = Mary   ($X = 20\%$ and $4X = 80\%$)
$4\,X$ = John

$4\,X + X$ = $180,000

$5\,X$ = 180,000

$$\frac{5\,X}{5} = \frac{180,000}{5} \qquad \frac{\cancel{5}\,X}{\cancel{5}} = \frac{180,000}{5} = 36,000$$

X = $ 36,000 Mary
$4\,X$ = $ 144,000 John

The above problem is the same as for the following. Joe is
one-fourth the age of his mother. Their combined ages
equal 55. How old is each of them?

X = Joe
$4\,X$ = Joe's mother

$4\,X + X$ = 55

$5\,X$ = 55

$$\frac{\cancel{5}\,X}{\cancel{5}} = \frac{55}{5}$$

X = 11 Joe
$4\,X$ = 44 Joe's mother

Word Problems
Mixtures

A collection of dimes and nickels equals $ 1.55. There are 20
coins. How many dimes and nickels are in the
collection?

$$X = \text{Dimes}$$
$$20 - X = \text{Nickels}$$

$$10¢\ X + 5¢(\ 20 - X\ ) = 1.55$$

remove the decimal point

$$10\ X + 100 - 5\ X = 155$$

$$10\ X + 100 - 100 - 5\ X = 155 - 100$$

$$10\ X - 5\ X = 55$$

$$5\ X = 55$$

$$\frac{\cancel{5}\ X}{\cancel{5}} = \frac{55}{5}$$

$$X = 11 \text{ Dimes}$$

$$20 - 11 = 9 \text{ Nickels}$$

There are 11 Dimes and 9 Nickels in the collection.

Word Problems

Mixtures

A bag of nuts costs $ 1.50. Nut A cost 42¢ an ounce, Nut B
cost 35¢ an ounce and Nut C cost 30¢ an ounce. How
many ounces of each nut are in the bag?

Total the cost of each nut.   .42 + .35 + .30 = 1.07

Divide this number into the cost of the bag.

$$\frac{1.50}{1.07} = 1.40187 \text{ ounces of each nut}$$
are in the bag.

Check
    1.40187 X .42 = 0.5887
    1.40187 X .35 = 0.4907
    1.40187 X .30 = 0.4206
                Total = 1.5063

A 32 ounce bottle has 3 chemicals to be mixed with water.
Chemical A is to be mixed 1-15 parts water, Chemical
B 1-7 and Chemical C is mixed 6 ounces to 1 gallon.
How many ounces of each are to be put into the bottle?

Chemical A $\frac{32}{15}$ = 2.13 ounces   Chemical B $\frac{32}{7}$ = 4.57 ounces

32 ounce = 1 quart   Chemical C $\frac{6}{4}$ = 1.5 ounces
4 quarts per gallon

To mix a chemical 1 part per 1,000

128 ounces = 1 gallon $\frac{128}{1,000}$ = 0.128 ounces

If using cubic centimeters   1 ounce = 29.57 cubic centimeters

        0.128 X 29.57 = 3.78 cubic centimeters

Word Problems
Mixtures

There are two groups of coins. The first group has 13 of one
kind of coin and 9 of another kind of coin. This group
equals $ 8.75. The second group has 15 of the first
kind and 11 of the second kind. This group equals
$ 10.25. What two kinds of coins are being used?

a = value of 1st kind
b = value of 2nd kind

remove the decimal points

1st group   13 a + 9 b = 875
2nd group   15 a + 11 b = 1,025

The b coins have to be canceled out, when one group is
subtracted from the other, so multiply each group by a
number that will cancel the b coins.

for the first group 11 X 9 = 99   the 2nd group   9 X 11 = 99

11 X 13 = 143   11 X 9 = 99   11 X 875 = 9625

$$11 \times (13a + 9b = 875) = 143a + 99b = 9625$$
$$9 \times (15a + 11b = 1025) = 135a + 99b = 9225$$
$$\frac{8a}{} = 400$$

Subtract the 2nd group from the 1st group

9 X 15 = 135   9 X 11 = 99   9 X 1025 = 9225

$$\frac{\cancel{8}a}{\cancel{8}} = \frac{400}{8} = 50 \quad a = 50 ¢$$
the first coin

Now work the equation for the first group of coins.

13 a + 9 b = 875   *We have found that a = 50*

13 X 50 + 9 b = 875

650 + 9 b = 875

650 – 650 + 9 b = 875 – 650   = 225

$$9b = 225 \quad \frac{9b}{9} = \frac{225}{9} = 25 \quad b = 25 ¢$$
the second coin

Check 13 X .50 + 9 X .25 = 8.75   6.50 + 2.25 = 8.75
Check 15 X .50 + 11 X .25 + 10.25   7.50 + 2.75 = 10.25

Word Problems
Mixtures

There are 3 bags of money, Bag A is $ 15.00 more than Bag B
and Bag C is $ 3.00 less than Bag A. The total value of
all 3 bags is $ 282.00. What is the value of each bag?

Bag A = X Bag B = $X - 15$ Bag C = $X - 3$

$$X + X - 15 + X - 3 = 282.00$$

$$3X - 15 - 3 = 282.00$$

$$3X - 18 = 282.00$$

$$3X - 18 + 18 = 282.00 + 18$$

$$3X = 300.00$$

$$\frac{3X}{3} = \frac{300.00}{3}$$

$$X = 100$$

Now work the equation with 100 replacing X.

$$\underset{\text{Bag A}}{100} + \underset{\text{Bag B}}{100 - 15} + \underset{\text{Bag C}}{100 - 3} = 282$$

$$100 + 85 + 97 = 282$$

Bag A = $ 100.00
Bag B = $ 85.00
Bag C = $ 97.00
          $ 282.00

Word Problems
Mixtures

A collection of half dollars, quarters, dimes and nickels equals $ 10.55. There are 47 coins. How many of each kind are there? (A simple formula for three or more unknowns does not exist, however a simple method does work very well.

Add the denomination of each coin.
50 + 25 + 10 + 5 = 90
Divide this amount into the total value of the coins.
$$\frac{10.55}{90} = 11.72$$
Ignore the partial portion (.72)

Now multiply each denomination by 11

$$
\begin{array}{rcl}
11 \times .50 &=& 5.50 \\
11 \times .25 &=& 2.75 \\
11 \times .10 &=& 1.10 \\
11 \times .05 &=& .55 \\
\end{array}
$$
Total the columns 44   9.90

Subtract from the required amount   47 coins   10.55
                                       44   9.90
                        Leaving   3 coins   .65

Using the selection of coins find 3 that equals .65, which are a half dollar, dime and a nickel, equaling 0.65 and the 3 additional coins.

half dollars	12 X 50	= 6.00
quarters	11 X 25	= 2.75
dimes	12 X 10	= 1.20
nickels	12 X .05	= .60
Total the columns	47	10.55

Mixtures

A cash drawer is to have close to an equal number of quarters, dimes, nickels and pennies. The cash drawer of coins, is to equal $ 20.00. How many of each coin will there be?

y = number of coins

$25¢y + 10¢y + 5¢y + 1¢y = \$ \ 20.00$

The total of 25+10+5+1 = 41   Remove the decimal
from $20.00 = 2,000 and divide by 41
2,000 divided by 41 = 48.79
Round off 48.79 to 49

.25 X 49 = 12.25
.10 X 49 =   4.90
.05 X 49 =   2.45
.01 X 49 =   .49
  total   20.09

Remove 1 dime or two nickels and you will have 19.99. Remove 1 nickel and 4 pennies and you will have the exact amount.

Word Problems
Multiple Task

Tony has completed 75% of a job. Carol then finished it in 3
   hours. The total time both spent on the job was 15
   hours. How long would it have taken Tony to finish the
   job?

Tony   Carol
0.75 + 3 = 15 hours

Eliminate the 3

0.75 +3 – 3 = 15 –3

0.75 = 12

Divide 12 by 0.75 equaling 16 hours for Tony
                            to have finished the job.

Check
0.75 of 16 + 3 = 15
   12 + 3 = 15

Tom and Joe can inspect 400 items a day on an assembly line
   working together. Tom works twice as fast as Joe.
   How many could the each inspect per day if they
   worked separately?

Tom = 2a   Joe = a

2a + a = 400

$\dfrac{3a}{3} = \dfrac{400}{3}$

a = 133.33

2a = 266.66

Tom can do 266.66 per day and
Joe can do 133.33 per day.

In the previous problem there were two elements, Tom and Joe. It is harder to identify elements when there are more than 2 . Also if you use a letter for one of the elements it is harder to work if you have to divide within an element. To eliminate confusion and enable problems to be structured beyond 3 or more elements it is easier to structure the problem in fraction form.

It takes Bob, Joe and Tom 5.5 days working together to paint a house. Bob works 2 times faster than Joe and Bob works 3 times faster than Tom. How long would it take for each to paint a house if they worked separately? Bob is the standard, Joe works 1/2 as fast as Bob and Tom works 1/3 as fast as Bob.

$$\text{Bob } \frac{1}{1} \qquad \text{Joe } \frac{1}{2} \qquad \text{Tom} \frac{1}{3}$$

Now set up the problem as if you were going to add fractions. Make all of the denominators the same number.

$$\text{Bob } \quad \frac{1}{1} \ \bigg|\ \frac{6}{6}$$
$$\text{Joe } \quad \frac{1}{2} \ \bigg|\ \frac{3}{6}$$
$$\text{Tom } \quad \frac{1}{3} \ \bigg|\ \frac{2}{6}$$

Now add up the numerators   6+3+2 = 11
Multiply this number times the amount of time for all three all to paint a house.

$$11 \times 5.5 = 60.50$$

Divide this number by each of the numerators.

$$\frac{60.50}{6} = 10.08 \qquad \frac{60.50}{3} = 20.166 \qquad \frac{60.50}{2} = 30.25$$

Working separately, it will take Bob 10.08 days, Joe 20.166 days and Tom 30.25 days to paint a house.

Word Problems
Multiple Tasks

Working separately Frank can paint a house in 7 days, Jack in 9 days and George in 11 days. How long will it take if they all work together on the same house?

Assign the value of "1" to the task, painting the house.

Frank = $\frac{1}{7}$ Jack = $\frac{1}{9}$ George = $\frac{1}{11}$

Now set the problem up as if you were going to add fractions.

$$\frac{1}{7} \left| \frac{99}{693} \right.$$

Multiply 7 X 9 X 11 = 693 for the new denominator.

$$\frac{1}{9} \left| \frac{77}{693} \right.$$

Figure the new numerators

$$\frac{1}{11} \left| \frac{63}{693} \right.$$

693 divided by 7 = 99.   693 divided by 9 = 77.   693 divided by 11 = 63.

Add these up 99 + 77 + 63 = 239

We will now depart on how fractions are handled in this form.

Now divide into the <u>denominator</u>

693 divided by 239 = 2.899 Days

If Frank, Jack and George work together it will take them 2.899 days to paint the house.

Another way of doing this problem.

$$\frac{1}{\dfrac{1}{7} + \dfrac{1}{9} + \dfrac{1}{11}}$$

$1 / 7$ = .14286
$1 / 9$ = .11111 — add
$1 / 11$ = .09091
.34488

1 divided by .34488 = 2.899 days

Word Problems

Time and Distance

Mary has to go 145 miles. If she averages 64 miles per-hour, how long will it take her?

$$\frac{\text{Distance}}{\text{Miles per-hour}} = \frac{145}{64} = 2.27 \text{ hours}$$

which is a little more than 2 hours and 15 minutes

Don has to go 32 miles. Traffic is moving at 62 miles per-hour on a 4 lane highway. If he switches lanes very often he can average 70 miles per-hour. How much sooner will he arrive at his destination?

First figure out how long it would take at the normal safe speed, then subtract Toms travel time.

$$\frac{32 \text{ miles}}{62 \text{ m.p.h.}} = 0.516 \text{ hours} \times 60 \text{ minutes} = 30.96 \text{ minutes}$$

$$\frac{32 \text{ miles}}{70 \text{ m.p.h.}} = 0.457 \text{ hours} \times 60 \text{ minutes} = 27.42 \text{ minutes}$$

$$\begin{array}{r} 30.96 \\ -27.42 \\ \hline 3.54 \end{array}$$ minutes saved by driving recklessly.

Mary is driving the family car and averages 70 m.p.h. and Joe is driving a truck and averages 60 m.p.h. Their next stop is 300 miles away. If they leave at the same time, how long will Mary be at their next stop before Joe gets there?

First figure out how long is will take each of them to get there.

$$\frac{300}{70} = 4.286 \text{ hours} \qquad \frac{300}{60} = 5 \text{ hours}$$

Then subtract one from another.

$$\begin{array}{r} 5.000 \text{ hours of Joe's time} \\ - 4.286 \text{ hours of Mary's time} \\ \hline .714 \times 60 \text{ minutes} = 42.86 \text{ minutes} \end{array}$$

Word Problems
Time and Distance

Carol is traveling at 65 miles per hour. She leaves 15 minutes
before Bob. How fast will Bob have to travel to catch
up with Carol in 3 hours?

First figure out how far Carol will have traveled.
15 minutes = .25 hours then add the 3 hours.
3.25 X miles per-hour Carol travels
3.25 X 65 = 211.24 miles

Next figure out how fast Bob will have to travel to
catch up with Carol.

$\dfrac{211.25}{3 \text{ hours}}$ = 70.417 m.p.h. Bob will have to average
to catch up will Carol in 3 hours.

Dave travels at 65 m.p.h. Sue travels at 42 m.p.h. If Sue leaves
18 minutes earlier than Dave, how long will it take Tom
to catch up will Sue?

First figure out miles per minute each will travel.

Dave = $\dfrac{65 \text{ mph}}{60 \text{ minutes}}$ = 1.0833 miles per minute

Sue = $\dfrac{42 \text{ mph}}{60 \text{ minutes}}$ = 0.70 miles per minute

$\left(\dfrac{\text{Lead Time}}{\underset{\text{miles per minute}}{\text{Fast Traveler}} - \underset{\text{miles per minute}}{\text{Slow Traveler}}}\right) = \dfrac{18 \text{ minutes}}{1.0833 - 0.70}$

$1.0833 - 0.70 = 0.38333$

$\dfrac{18 \text{ minutes}}{0.3833}$ = 46.96 minutes

Proof- Sue 18 min. lead time + (46.96 minutes X 0.70 miles per minute)
18 + 32.872 = 50.87 miles
Dave 46.96 minutes X 1.0833 miles per minute = 50.87

Dave and Sue will meet up 46.96 minutes after Dave
leaves and both will have traveled 50.87 miles.

Sales Tax

Price + Sales Tax = Total
Price X Sales Tax Rate = Sales Tax
Example
    Price is $24.50
    Sales tax rate is .06 (6%)
    24.50 X .06 = 1.47
    24.50 + 1.47 = 25.97 the total
To find the total without needing the amount of tax, simply multiply the Price by 1 and the sales tax rate.
Example
    Price is $36.99
    Sales tax rate is .0825
    36.99 X 1.0825 = $40.04 gives the total of the transaction.

Backing Out Sales Tax or Price

If you have the total amount of a transaction and you need to know what the price of the item is, or the amount of the sales tax, do the following. Divide the total by the sales tax rate plus 1, will give you the price. You can then figure out the amount of tax.

Example- Price unknown
    Sales tax rate is 8%
    Total is $38.60

$$\frac{38.60}{1.08} = 35.74 \text{ cost of item}$$

38.60 - 35.74 = 2.86 tax

Price = 35.74
Tax =   2.86
Total = 38.60

Business - Sales

Markup Based on Cost- This is used when you want to markup your cost, based on a certain percent to arrive at a retail price.

Selling Price = Cost X (1 + Markup)
Example Cost = 24.50
Markup percent = 66.7 % has been selected

Selling Price = 24.50 X (1+.667)
= 24.50 X 1.667 = 40.84
Selling Price = $ 40.84

Markup Based on Selling Price- This is used when you want your profit to be a certain percent of the retail price.

$$\text{Selling price} = \frac{\text{Cost}}{1 - \text{Markup}}$$

Example #1 If you want 40% of the selling price to be profit and your cost is $ 24.50

$$\text{Selling Price} = \frac{24.50}{1 - .40}$$

$$\text{Selling Price} = \frac{24.50}{.60} = 40.83$$

Example # 2 How to find the retail price of an item, when your supplier gives you a discount of 25% off the retail price, resulting in your discounted cost being $48.00 .

$$\frac{\text{Retail}}{\text{Price}} = \frac{\text{Cost}}{1 - \text{discount}}$$

$$\frac{\text{Retail}}{\text{Price}} = \frac{48.00}{1.00 - .25}$$

$$\frac{48.00}{.75} = \$ 64.00 \text{ the retail price}$$

To check this
64 X . 25 = 16   64 - 16 = $48 your cost

Mark Up Based on Cost

If you want to multiply your cost, instead of using the dividing method, do the following.

Example- If you want your profit to be 25 % of the retail price.

Subtract the amount of, percent of profit, from 100

100 - 25 = 75

$\frac{25}{75}$ = 0.333

Now add the amount of 1 to 0.333 = 1.333

1.333 X 48 = 63.99 or $ 64

Example- If you want your gross margin (profit) to be 40%.

100 - 40 = 60

$\frac{40}{60}$ = .6667 and add 1

1.6667 X 24.50 = 40.83

Markup with a Load (usually sales commissions) on Selling Price

$$\text{Selling Price} = \frac{\text{Cost}}{1 - (\text{Markup} + \text{Load})}$$

Example- Cost of item = $ 65.00

Markup = 40%

Sales Commission = 15%

$$\text{Selling Price} = \frac{\$65}{1 - (.40 + .15)}$$

$$\frac{\$65}{1 - .55}$$

$$\frac{\$65}{.45} = \$144.44$$

To check answer

144.44 X .15 = $ 21.67 Sales Commission

144.44 X .40 = $57.78

21.67 + 57.78 = 79.45

144.44 - 79.45 = 64.99 or $65.00

Business - Sales

Markup with a Load on Cost

$$\text{Selling Price} = \frac{\text{Cost X (1 + Markup percent)}}{1 - \text{Load Factor percent}}$$

Example
Cost = $36.40
Markup percent = 67%
Load Factor = 15%
Selling Price = ??

$$\frac{36.40 \ (\ 1 + .67)}{1 - .15}$$

$$\frac{36.40 \text{ X } 1.67}{.85}$$

$$\frac{60.79}{.85} = 71.52 \text{ the selling price}$$

Break Even Point- The number of units sold above the break even point starts contributing to profit. You have to determine the number of units that you have to sell to cover the fixed cost and the variable cost.

$$\frac{\text{Break Even}}{\text{Units}} = \frac{\text{Fixed Cost}}{\text{Selling Cost - Variable cost per unit}}$$

Example- A book cost you $ 2,800 to advertise (fixed cost) and the printing and handling is $ 8.50 per unit (variable cost). How many books have to be sold to break even, if the book sells for $ 21.95?

$$\frac{\text{Break Even}}{\text{Units}} = \frac{\$2800}{\$21.95 - 8.50}$$

$$\frac{\text{Break Even}}{\text{Units}} = \frac{2800}{13.35} = 208.18 \text{ units}$$

Break Even Units = 209

Maintenance of Sales Dollar Verses Discount Percent

When a discount is given, you have to increase
sales to have the same dollar amount in sales
after the discount is given.

Example
There is $ 12,000 in previous sales, if a
discount of 15% is given, how much in
increased total sales will need to be made to
still have $12,000 in sales after the discount is
given?

Subtract the discount from 1, then divide that
number into the discount, equaling the percent
amount of increased sales, needed to maintain
the same total sales dollar amount.

1 - .15 = .85

$$\frac{.15}{.85} = .1764706$$

0.1764706 X 12,000 = $2,117.65
Add 2,117.65 to 12,000.00 = 14,117.65 in sales
to equal 12,000.00 after a discount of 15% is
given.
Check;

14,117.65 X .15 = 2,117.65
1,4117.65 - 2,117.65= $12,000.00

Discounts

Noncumulative = The base for bracket pricing. For example-
3 units for $ 2.40 each
4 - 10 units for $ 2.30 each
11 or more for $2.10 each

Cumulative = Usually expressed in a percentage off the
regular selling price.
Example - 30% off.

Cash Discounts or
Sales Policy or
Sales Terms = The amount the seller will discount from the
total bill, if the person paying the bill will pay
within a specific time. For example - 2/10 N/30
means the bill will be discounted 2 % if paid
within 10 days and the full amount to be paid
within 30 days.
Cost of Not Taking the Cash Discount - Using the above you
would base the cost on 20 days. 360 days
divided by 20 = 18. 18 X 2% = 36 %

Time of Payment
E. O. M. - End of Month
M. O. M. - Middle of Month
R. O. G. - Receipt of Goods

Accounts Receivable - Carrying Charges without
Compounding
If you do not want to compound the carrying charges,
you would do the following.

Amount of bill X 1 + {(rate per-month X number of months)}

938.50 X 1 +{(0.015 X 3)} = 938.50 X 1.045 =$ 980.73

Example- A customer has an unpaid balance on the
first of the month of $ 1,925.37. The carrying
charge is 1 1/2 % per-month. What is the total
bill?
1,925.37 X 0.015 = 28.88

1,925.37 + 28.88 = $ 1954.25 the customer now owes

To save time do the following instead

$ 1,925.37 X 1.015 = $ 1954.25

Most businesses charge 1 1/2 % per-month to their
customers on unpaid balances.

Example - A customer has not paid his bill of $ 938.50
for 3 months and finally you have decided to
charge him carrying charges of 1 1/2 % for each
month. The following formula will compound
the interest (carrying charges).
Amount of bill X (rate per-month + 1) $^{\text{number of months}}$

$$938.50 \text{ X } (0.015 + 1)^3$$

using a calculator 1.015 $\boxed{Y^x \text{ key}}$ 3 $\boxed{=}$ 1.0456784

this is the same as 1.015 X 1.015 X 1.015 = 1.0456784

938.50 X 1.0456784 = 981.37

Accounts Receivable - Month to Month

On the day of billing all payments that have came in within the current billing period gets credited, then the carrying charges are billed.

Example - The carrying charges of XYZ company is 1.5 % and is charged on the outstanding balance on the billing date.

Jan 10	1,250.75	Last billing
Feb 9	950.00	Received on acct.
Feb 9	300.75	Balance
Feb 10	305.26	(300.75 X 1.015)

Example - A customer buys a machine for $ 10,000 and agrees to make monthly payments of $ 2,000 per month at 15% annual carrying charges.

$$\frac{\text{annual rate}}{\text{months in the year}} \quad \frac{.15}{12} = 0.0125 \quad \frac{\text{monthly}}{\text{rate}}$$

	10,000.00	X	1.0125	=	10,125.00
1	10,125.00	-	2,000	=	8,125.00
	8,125.00	X	1.0125	=	8,226.56
2	8,226.56	-	2,000	=	6,226.56
	6,226.56	X	1.0125	=	6,304.39
3	6,304.39	-	2,000	=	4,304.39
	4,304.39	X	1.0125	=	4,358.20
4	4,358.20	-	2,000	=	2,358.20
	2,358.20	X	1.0125	=	2,387.68
5	2,387.69	-	2,000	=	387.68

The remaining amount, $ 387.68 is the interest that has accumulated during the last 5 months.

Please refer to the Money Chapter for more formulas, espically pages 9:10-12

The following text is only an introduction of how the government treats depreciation. There are many useful publications the IRS has printed that more thoroughly explains the program. However the following, will make theirs easier to understand.

The Internal Revenue Service has a very specific way of doing depreciation called the "Modified Accelerated Cost Recovery System," (MACRS). It consist of two systems, the General Depreciation System (GDS) and the Alternative Depreciation System (ADS). Depreciation done by Units of Production does not use either system and is treated as a special system. Each item you depreciate has a specific number of years you can depreciate. There are tables that outline the number of years for specific items and the amount that you can depreciate for each year. These are available in IRS Publication 946. You can also figure the percent of depreciation yourself. The following information will give you a reference as to how the system works.

General Depreciation System
The most widely used system for business. It has recovery times of 3,5,7,10,15 and 20 years. You can use 200%, 150% Declining Balance or Straight line methods.

200% Declining Balance Method (This is a text book illustration. Please refer to the example shown after **Conventions,** to see how IRS uses this method).

Example- Cost of item - $ 20,000 Deprecation - life 5 years
    Step 1 Determine the rate of depreciation based on its expected life. Each year would be a percentage of 200%. (200% = 2.00) 2 divided by 5 yrs = 0.40
    Step 2 20,000 X .40 = 8,000 deprecation 1 year
        20,000 - 8,000 = 12,000 2nd year value
        12,000 X .40 = 4,800 deprecation 2 year
        12,000 - 4,800 = 7,200 3rd year value
        7,200 X .40 = 2,880 deprecation 3 year
        7,200 - 2,880 = 4,320 4th year value
        4,320 X .40 = 1,728 depreciation 4 year
        4,320 - 1,728 = 2,592 5th year value
        2,592 X .40 = 1,036.80 deprecation 5 yr.
        2,592 - 1,036.80 = 1,555.20 remaining amount that will be claimed on the 6th year.

200% Declining Balance Method (continued)

Year	Cost	Book value at beginning of year		Rate		Depreciation for year
1	20,000	20,000	X	.40	=	8,000
2	20,000	12,000	X	.40	=	4,800
3	20,000	7,200	X	.40	=	2,880
4	20,000	4,320	X	.40	=	1,728
5	20,000	2,592	X	.40	=	1,036.80
6	20,000	1,555.20				1,555.20

Total 20,000.00

150% Declining Balance Method

Divide 1.50 by 5 = 0.30 Then do as what was illustrated in the above example.

Straight Line Method (SL)
Generally used for real estate and for longer recovery periods. It can be used under the General Depreciation System or the Alternative Depreciation System. When using the Alternative Depreciation System only the Straight Line Method can be used. The Straight line System is easier to use but does not allow for higher recovery soon after the item is placed in use.

Example- Cost of item - $ 20,000 Deprecation - life 5 years
Step 1 Determine the rate of depreciation based on its expected life. Each year would be a percentage of 100%. (100% = 1.00)
1 divided by 5 years = 0.20 for each year
Step 2 Multiply $ 20,000 X 0.20 = $ 4,000 for each year of depreciation.

The Straight Line Method is easy to work, however as IRS shows how to do it, it requires more steps. Their way becomes very useful when a partial year convention is used. (Conventions will be outlined later). The following illustrates the IRS way of doing the Straight line method.

Straight Line Method (continued)

IRS Procedure (the following is necessary when Conventions
are used and will be explained in the next
section.

Step 1 Divide each remaining year of the depreciation
period into 1. For a 5 year depreciation period
The first year would be 1 divided by 5 = 0.20;
the second year 1 divided by 4 = 0.25; the third
year 1 divided by 3 = 0.3333; the fourth year
1 divided by 2 = 0.5 and the last year would be
1.0

Step 2 Each year you subtract the depreciation from
that years beginning value.

Example- $ 20,000 depreciated for 5 years using SL

20,000 X 0.20 = **4,000**   ⟵
20,000 - 4,000 = 16,000
16,000 X 0.25 = **4,000**   ⟵
16,000 - 4,000 = 12,000
12,000 X 0.3333 = **4,000**   ⟵
12,000 - 4,000 = 8,000
8,000 X 0.50 = **4,000**   ⟵
8,000 - 4,000 = 4,000
4,000 X 1.0 = **4,000**   ⟵
4,000 - 4,000 = 0

Conventions- (the following is a very brief explanation)
Half-Year is used for all property that is not real
estate income property. Usually used when
more than one piece of property is depreciated.
All are counted as if they were started in the
middle of the year.
Mid-Quarter is used as long as the piece of property
is more than 40% of the equipment you are
depreciating. Counted as if started in the middle
of the quarter
Mid-Month usually used for income producing real
estate. Counted as if rented in the middle of the
month even if it was on the 1st or 30th day.

Business - Depreciation

Half-Year Convention

Example- Cost of property is $ 20,000 and the depreciation life is 5 years. Maximum depreciation is desired at the beginning, so the 200% Declining Balance Method will be used. However you will see that the Straight Line Method will also be used at the last, because you can use it, if it will allow for more depreciation.

Step 1 Determine the percent of depreciation for the first year. 200% = 2.0   2 divided by 5 = .40 With the Declining Balance Half-Year Convention you divide the first year in half, 0.40/2 = 0.20 On the right side we will check to see if the Straight Line will give higher depreciation Using Straight Line half-year Convention you take one-half of 1/5 (1/2 of 0.20 = 0.10) for the first year.

Declining Method	Straight Line Method
1st yr 20,000 X 0.20 = **4,000**	20,000 X 0.10 = 2,000
20,000 - 4,000 = 16,000	there is now 4.5 yrs remaining
2nd yr 16,000 X 0.40 = **6,400**	16,000 X 1/4.5 = 3,556
16,000 - 6,400 = 9,600	3.5 remaining years
3rd yr 9,600 X 0.40 = **3,840**	9,600 X 1/3.5 = 2,743
9,600 - 3,840 = 5,760	2.5 remaining years
4th yr 5,760 X 0.40 = **2,304**	5,760 X 1/2.5 = **2304**

Please note, both methods have the same results for the 4th yr.

5,760 - 2,304 = 3,456	1.5 remaining years
5th yr 3,456 X 0.40 = 1,382	3,456 X 1/1.5 = **2,304**

The Straight Line deduction is taken because it is higher than the Declining Balance Method.

6th yr   Any remaining depreciation is claimed

3,456 - 2,304 = **1,152**

Business - Depreciation

Example chart for Conventions
The following chart is one of 18 charts for the 3
different Conventions and different depreciation
methods. It is the chart for most business deductions.

Depreciation rate for recovery period

Year	3 year	5 year	7 year	10 year	15 year	20 year
1	33.33%	20.00%	14.28%	10.00%	5.00%	3.750%
2	44.45	32.00	24.49	18.00	9.50	7.219
3	14.81	19.20	17.49	14.40	8.55	6.677
4	7.41	11.52	12.49	11.52	7.70	6.177
5		11.52	8.93	9.22	6.93	5.713
6		5.76	8.92	7.37	6.23	5.285
7			8.93	6.55	5.90	4.888
8			4.46	6.55	5.90	4.522
9				6.56	5.91	4.462
10				6.55	5.90	4.461
11				3.28	5.91	4.462
12					5.90	4.461
13					5.91	4.462
14					5.90	4.461
15					5.91	4.462
16					2.95	4.461
17						4.462
18						4.461
19						4.462
20						4.461
21						2.231

Units of Production Method
Original Cost, divided by the number of any useful
unit of production, multiplied by the number of items
made during the depreciation period.

Example- A cylinder that cost $15,000 can produce
2,000,000 items. During the year 425,00
items were made.

$$\frac{\$15,000}{2,000,000} = 0.0075 \text{ depreciation per item}$$

$ 0.0075 X 425,000 = $ 3,187.50 Depreciation for
the year.

Business - Ratios

For the following ratios we will use information from XYZ Company.

Gross Sales -	165,000	Annual Credit Sales	148,000
Sales Returns and allowances	750	Accounts Receivables	13,200
Sales Discounts	2,200	Average Account 5 days past due	
Net Sales	162,050	Average Accts Receivables	13,600
Cost of Goods Sold	61,000	Cash	3,800
Gross Profit on Sales	101,050	Security Bonds	7,000
Operating Expenses	65,000	Current Assets	58,100
Net Operating Income	36,050	Fixed Assets	52,100
Other Revenue	200	Total Assets	110,200
Other Expenses	2,100	Current Liabilities	22,000
Net Income Before Taxes	34,150	Bonds Payable	25,000
Provision for Income Taxes	8,900	Average Inventory	16,760
Net Income after Taxes	25,250	Retained Earnings	36,000
		Stock Holders Equity	76,000
		Common Stock	25,000
		Preferred Stock	15,000

Common Stock issued & outstanding
(par $ 5.00) 5,000
Preferred Stock issued & outstanding
(par $ 50.00 8.25%) 300
Common Dividend
($ 0.12 per stock) $ 600.00

Acid Test (Quick Ratio)

$$\frac{\text{Cash + Marketable Securities + Accounts Receivables}}{\text{Current Liabilities}} =$$

$$\frac{3,800 + 7,000 + 13,200}{22,000} = \frac{24,000}{22,000} = 1.09 \text{ to } 1$$

1.00 or higher is standard

Current Ratio

$$\frac{\text{Current Assets}}{\text{Current Liabilities}} = \frac{58,100}{22,000} = 2.64 \text{ to } 1$$

2.0 or higher is standard

Cash to Debt

$$\frac{\text{Cash}}{\text{Total Debt}} = \frac{3,800}{27,000} = 14\ \%$$

Accounts Receivable Turnover

$$\frac{\text{Annual Credit Sales}}{\text{Average Accounts Receivable}} = \frac{148,000}{13,600} = 10.88 \text{ to } 1$$

Accounts Receivable to Sales Year

$$\frac{\text{Annual Credit Sales}}{\text{Net Sales}} = \frac{148,000}{162,050} = .91 \text{ to } 1$$

Sales to Current Assets

$$\frac{\text{Sales}}{\text{Current Assets}} = \frac{165,000}{58,100} = 2.84 \text{ to } 1$$

Total Asset Turnover

$$\frac{\text{Sales}}{\text{Total Assets}} = \frac{165,000}{110,200} = 1.5 \text{ to } 1$$

Investment in Accounts Receivable

$$\frac{\text{Days Held}}{360} \text{ X Annual Credit Sales}$$

$$\frac{30 + 5}{360} \text{ X } 148,000 = \frac{35}{360} \text{ X } 148,000$$

$$.0972 \text{ X } 148,000 = 14,385.60$$

Inventory Turnover

$$\frac{\text{Cost of Goods Sold}}{\text{Average Inventory}} = \frac{61,000}{16,760} = 3.64 \text{ to } 1$$

Age of Inventory

$$\frac{365}{\text{Inventory Turnover}} = \frac{365}{3.64} = 100.27 \text{ Days}$$

Business - Ratios

Sales to Inventory

$$\frac{\text{Sales}}{\text{Inventory}} = \frac{165,000}{16,760} = 9.8 \text{ to } 1$$

Fixed Asset Turnover

$$\frac{\text{Sales}}{\text{Fixed Assets}} = \frac{165,000}{52,100} = 3.14 \text{ to } 1$$

Return on Total Assets

$$\frac{\text{Net Income}}{\text{Average Total Assets}} = \frac{34,150}{110,200} = 0.31 \text{ to } 1$$

Fixed Assets to Stockholders Equity

$$\frac{\text{Fixed Assets}}{\text{Stockholders Equity}} = \frac{52,100}{76,000} = 0.69 \text{ to } 1$$

Net Profit Margin

$$\frac{\text{Net Income (after Taxes)}}{\text{Net Sales (before Cost of Goods Sold)}} = \frac{25,250}{162,050} = 0.16 \quad 16\%$$

Gross Profit Margin

$$\frac{\text{Gross Profit (after Cost of Good Sold)}}{\text{Net Sales (before Cost of Goods Sold)}} = \frac{101,050}{162,050} = 0.62 \quad 62\%$$

Retained Earnings to Stockholders Equity

$$\frac{\text{Retained Earnings}}{\text{Stockholders Equity}} = \frac{36,000}{76,000} = 0.47 \text{ to } 1$$

Equity Growth Rate

$$\frac{\text{Net Income (minus) Preferred Div. \& Common Dividends}}{\text{Average Common Stockholders Equity}}$$

Preferred Stock of $15,000 X .0825 = $ 1237.50

$$\frac{25,250 \text{ (minus) } 1237.50 \,\&\, \$600}{76,000} = \frac{23412.50}{76,000} = .308$$

30.8 %

Simple Rate of Return

The annual rate of return on an investment. For example, if you buy equipment that cost $ 28,000 and make net income from it of $ 7,000 per year.

$$\frac{\text{Expected Future Annual Net Income}}{\text{Average Investment}} = \frac{7,000}{28,000} = .25$$

25 % Rate of return

Return on Investment (ROI)

$$\frac{\text{Investment}}{\text{Annual Net Income}} = \frac{28,000}{7,000} = 4 \text{ years}$$

Earnings Per Share

$$\frac{\text{Net Income}}{\text{Total Shares}} = \frac{\$ 25,250}{5,000} = \$ 5.05$$

Business - Investments

Yield on Taxable verses Tax-Free Investments

When making the decision between a tax free or taxable investment you need to know at what rate you need to have, to obtain the same yield after taxes.

Example- You have an opportunity to invest $40,000 for 20 years at a rate of 5% in a tax free bond. If your tax rate is at 31% at what rate do you need to invest at, in a taxable bond to have the same rate of return?

$$\text{Taxable equivalent yield} = \frac{\text{Rate of Tax free Yield}}{1 \text{ minus your tax rate}}$$

$$\text{Taxable equivalent yield} = \frac{5.0}{1 - .31}$$

$$\text{Taxable equivalent yield} = \frac{5.0}{0.69} = 7.246$$

$$\text{Taxable equivalent yield} = 7.25 \%$$

You would have to invest at a rate of 7.25% in a taxable investment to yield the same amount that you would yield from a tax free investment with a rate of 5 %.

Proof-

$40,000 X .0725 = 2,900

$2,900 is taxed at a rate of 0.31 = $899.00

2,900 - 899 = 2,001 which is about the same (+$1) if you would have invested in a tax-free bond.

Price Earnings Ratio (P/ E Ratio)

When the price of each share of stock is divided by the annual earnings of each share of stock, the answer is the price earnings ratio.

$$\frac{\text{Stock Price Per Share}}{\text{Annual Earnings Per Share}} = \text{Price Earnings Ratio}$$

Example- The price of a share of stock is selling for $ 87.50. The company reports that earnings per share for the year is $ 6.25.

$$\frac{\$\ 87.50}{\$\ 6.25} = 14\ \text{P/E\ Ratio}$$

Yield - Return

The yield or return on your money is usually expressed as a percent based on the amount you have invested.

Example - You bought 500 shares at $ 78.50 per share. A year later you sell the 500 shares for $ 91.00 each.

Original investment 500 X $ 78.50 = $ 39,250
Current investment value 500 X $ 91.00 = $ 45,500

$$\text{Yield} = \frac{\text{Current investment value - Original Investment}}{\text{Original Investment}}$$

$$\text{Yield} = \frac{45,500 - 39,250}{39,250} = \frac{6,250}{39,250} = 15.9\ \%$$

Business - Investments

Total Return

> **Example -** Using the previous example we will now
> figure in the brokers fee for buying and selling
> the stock plus the dividend that was paid.
>
> A dividend of $5.50 for each share was paid.
> $5.50 X 500 = $ 2,750
> A brokers fee $ 30.00 was charged for buying
> and another $ 30.00 for selling the stock.

$$\frac{\text{Current Investment Value} - \text{Original Investment} + \text{Dividend} - \text{Brokers Fee}}{\text{Original Investment}}$$

$$\frac{45,500 - 39,250 + 2,750 - 60.00}{39,250}$$

$$\frac{8,940}{39,250} = 22.77 \text{ \% Total Return}$$

Bonds - Current Yield

$$\text{Current Yield} = \frac{\text{Coupon interest payment}}{\text{Market Price of the Bond}}$$

Example- A 20 year Bond paying 5.25% has a face value of $ 1,000.00 . It is purchased for $ 1,020.00. What is the current yield?

$$\text{Current Yield} = \frac{0.0525 \text{ X } 1,000}{1,020}$$

$$\text{Current Yield} = \frac{52.50}{1,020} = 0.0514706 = 5.147\%$$

Bonds - Yield to Maturity (approximate)

There are many formulas for determining Yield to Maturity of a Bond, either actual or approximate. The formulas for actual yield are very involved and will not be discussed in this text. Their use does not reflect a higher degree of accuracy that makes it worth the effort to work a far more involved formula. The following formula for approximating the yield will also give the values for average income, and average cost.

The above bond example will be used

$$\frac{\text{Yield to}}{\text{Maturity}} = \frac{\text{Average income}}{\text{Average cost}}$$

$$\text{Average income} = \left(\text{Interest X } \frac{\text{Face}}{\text{Value}}\right) - \frac{\text{Paid} - \text{Face}}{\text{Amount} - \text{Value}}{\text{Number of Years}}{\text{to Maturity}}$$

$$\text{Average income} = (\text{ } 0.0525 \text{ X } 1000 \text{ }) - \frac{1020 - 1000}{20}$$

$$\text{Average income} = 52.5 - \frac{20}{20} = 52.5 - 1 = 51.5$$

Average income is $ 51.50

(*Continued on next page*)

Bonds - Yield to Maturity (approximate) *continued*

$$\text{Average Cost} = \frac{\text{Cost of Bond} + \text{Maturity Value}}{2}$$

$$\text{Average Cost} = \frac{1,020 + 1,000}{2} = \frac{2,020}{2} = 1,010$$

Average Cost = $1,010.00

$$\text{Approximate Yield} = \frac{\text{Average Income}}{\text{Average Cost}}$$

$$\text{Approximate Yield} = \frac{51.50}{1,010} = 0.509901$$

Approximate Yield = 5.1 %

Bonds - Payment of Earnings

The interest paid from Bonds starts not from the day they are bought but from the day they are issued, called the "Day of Record." If the coupon payment (interest) is to be paid twice a year, in July and December and the owner of record is June 30th it will be paid to that person even if they just bought it.

Example- A person buys a $1,000.00 bond on June 10th for $950.00 The coupon rate is 7.5 % and is paid twice a year, in July and January. How much will they receive after July 1st.

$$\underset{\text{times a year}}{\frac{0.075}{2}} = 0.0375$$

$$0.0375 \times 1,000 = \$37.50$$

Money - Currency Exchange

Currently there are about 56 different countries that have their currency traded around the world. Listed in most newspapers is the rate of exchange. It can change daily and on which side of the border you are on.

Partial List	Foreign Currency in Dollars	Dollar in Foreign Currency
Australia (Dollar)	.6295	1.5886
Belgium (Franc)	.0244	40.98
Britain (Pound)	1.5991	.6254
Canada (Dollar)	.6894	1.4505
France(Franc)	.1501	6.6601
Germany Mark)	.5036	1.9858
Italy (Lira)	.000509	1958.6
Japan (Yen)	.00902	109.99
Spain (Peseta)	.005901	169.45
Russia (Ruble)	.0348	28.74
Venezuela (Bolivar)	.0015	659.5

Example- If you are in Australia and you want to change your U.S. Dollars to Australian Dollars

1 U.S. Dollar = 1.5886 Australian Dollars
Multiply your U.S. Dollar by 1.5886 to get Australian $
U.S. 63 cents = 1 Australian Dollar
When you leave and exchange Australian Dollars to U.S. Dollars
1 Australian Dollar = .63 cents
Multiply your Australian Dollars by 0.63 to get U.S.$

In Russia   U.S. 3 1/2 cents =1 Ruble or 1 U.S. Dollar = 28.74 Rubles

When going from Japan to Venezuela it works best to change Yen to American dollars, then to Bolivar.

1,000 Yen = $9.02 U.S. Dollars = 9.02 X 659.5 = 5,948.69 Bolivars

Please note, the average 14 year old street vendor, selling inexpensive jewelry in a foreign country, knows more about currency exchange rates, than anyone in your local bank.

Money - Simple Interest

Ordinary Simple Interest

Interest = Principal X Rate X Time

Example- Principal = $4,500
Interest Rate = .06
Time = 120 days
How much will the interest be?

$$4,500 \text{ X } .06 \text{ X } \frac{120}{360}$$

120 days is a portion of the year and simple interest is calculated using 360 days for the year.

$$\frac{120}{360} = .3333$$

4,500 X .06 X .3333
4,500 X .06 = 270
270 X .3333 = $89.99 the amount of

Exact Simple Interest

interest.

You use the same formula as above however you use 365 days instead of 360.

Time Needed to Yield a Determined Amount Using Simple Interest

$$\text{Time} = \frac{\text{Interest}}{\text{Principal X Rate}}$$

Example- How long do you have to leave $6,000 in the bank, to make a total of $7,000 at simple interest at 5%.
Time = ??
Principal is $6,000
Interest = $7,000 - $6,000 = 1,000
Rate = .05

$$\text{Time} = \frac{1,000}{6,000 \text{ X } .05}$$

$$\text{Time} = \frac{1,000}{300}$$

Time = 3.33 years

Loan Amount Needed to Obtain Determined Amount of Cash -

Simple Interest Loan

Principal = Cash X 1+(rate X 2)

Example　Principal = ??
Cash you want = $ 8,500
Rate = 0.07 annual rate
Time = 2 year loan period

Principal = 8,500 X 1 (0.07 X 2)

Principal = 8,500 X 1.14 = $ 9690

Principal = $ 9,690.00 the amount of money you
would have to borrow to walk
out with $ 8,500

Compound Interest (not simple interest)

Principal = Cash X (1 + rate)^{Time}

Example - Using the above example
Principal = 8,500 X (1 + 0.07)^{2}

1.07 X 1.07 = 1.1449

Using a calculator do the following
1.07 $\boxed{Y^X \text{ Key}}$ 2 $\boxed{=}$ 1.1449 X 8,500 = 9,731.65

Principal = 9,731.65 the amount of money you would
have to borrow to walk out with
$ 8,500

Discount Method of Interest is what is illustrated in the
above examples. It is when you the borrower , pay the
interest at the beginning of the loan period. The
amount of interest is deducted from the amount loaned
and the remaining is what you receive from taking out a
loan.

Money - Interest

Annual Percentage Rate-

When you get a loan you are provided two rates, the stated rate and the annual percentage rate, also called the annual effective interest rate. To best explain the difference, let us look at depositing money into the bank, for a moment.

If you deposit 500 for one year at 6 % interest and the interest is only compounded annually, you would have

$500 X 1.06 = $530 at the end of one year

However if the bank pays interest daily you would have more money at the end of one year because interest would be compounded daily. Compounding daily would give you $530.92 which is 92 cents more. This increased amount equals interest of 6.183 %.

Likewise when you borrow money, the loan company charges against the unpaid principal, the stated rate of interest compounded daily. This then becomes the effective interest rate. What can also bring up the effective interest rate is the financing of any loan fees.
When a bank loans money they usually charge a monthly rate against the unpaid balance. To find this rate do the following.

Divide the interest rate by 365.
Add 1 to it then multiply it times itself, 30 times.

Example : What is the monthly rate charged against the unpaid balance of a 35 % rate of interest when the rate is compounded daily?

$$\frac{.35}{365} + 1 \text{ times itself 30 times}$$

Also done mathematically

$$\left(1 + \frac{.35}{365}\right)^{30}$$

(continued)

Annual Percentage Rate *(continued)*

On a calculator do the following

.35 ÷ 365 = .000958904 + 1 = 1.000958904

1.000958904 Y^X 30 = 1.029170708 monthly rate
Days

Example: $1,000 is borrowed for 1 year at 35 % interest.
If a loan payment is not made until after 30 days , how
much is charged against the loan for the first 30 days?

1,000 X 1.029170708 = 1,029.17

After the first payment of $ 100 is made , how much is
charged against the loan?

$ 1,029.17 - 100 = 929.17 X 1.02917070 = $ 956.28

This is what the complete loan/payment program looks like per
month.

		monthly interest			Balance after payment
1,000 X .029170708 =	29.17 + 1,000	= 1,029.17 -	100	= 929.17
929.17 X .029170708 =	27.10 + 929.17	= 956.27 -	100	= 856.27
856.27 X .029170708 =	24.98 + 856.27	= 881.25 -	100	= 781.25
781-25 X .029170708 =	22.79 + 781.25	= 804.04 -	100	= 704.04
704.04 X .029170708 =	20.54 + 704.04	= 724.58 -	100	= 624.58
624.58 X .029170708 =	18.22 + 624.58	= 642.80 -	100	= 542.80
542.80 X .029170708 =	15.83 + 542.80	= 558.63 -	100	= 458.63
458.63 X .029170708 =	13.38 + 458.63	= 472.01 -	100	= 372.01
372.01 X .029170708 =	10.85 + 372.01	= 382.85 -	100	= 282.86
282.56 X .029170708 =	8.24 + 282.56	= 290.08 -	100	= 190.80
190.80 X .029170708 =	5.57 + 190.80	= 196.37 -	100	= 96.37
96.37 X .029170708 =	2.81 + 96.37	= 99.18 -	100	= -.82

Please note that due to rounding off, the totals and end results do not zero out.

The total amount paid out was $1200.00. The amount of
interest paid was $ 199.48. This means that the cost of the loan
was about 20 % even though it was based on 35%. This is
because the loan was paid on during the year.

APR - Annual Percentage Rate (continued)

To determine the APR of a Loan

$\quad$ M = number of payments in a year
$\quad$ N = number of planned payments
$\quad$ C = cost of loan in dollars
$\quad$ P = original loan proceeds

$$APR = \frac{M\,(\,95N + 9\,)\,C}{12N\,(\,N + 1\,)\,[(\,4P + C\,)\,]}$$

Example - A discounted loan of $ 20,000 with 48 monthly payments of $ 556.62 means that $ 2,6717.76 will have to be paid back. This means that the cost of the loan is $ 6,717.76 What is the APR?

$$APR = \frac{12\,(\,95\times48+9\,)\times6,717.76}{12\times48\times(\,48+1\,)\times[4\,(\,20,000\,)+6,717.76]}$$

$$APR = \frac{12\times(\,4560+9\,)\times6,717.76}{12\times48\times49\times[80,000\ +\ 6,717.76]}$$

$$APR = \frac{12\times4,569\times6,717.76}{12\times48\times49\times8,6717.76}$$

$$APR = \frac{368,321,345.3}{2447522058} = 0.15048$$

$$APR = 15.048\ \%$$

Determining Interest Rate

You can find the unknown rate of interest when you know the total amount due, the original amount and the amount of time that has elapsed.

Example - Interest Compounded Yearly

The original amount was $ 5,000 and no payments had been made for 3 years. After 3 years the total amount due was $ **7,214.49** . What was the rate of interest if the interest was compounded **yearly**?

The amount of interest is

$$7,214.49 - 5,000 = 2,214.49$$

The amount of interest is then divided by the original amount.

$$\frac{2,214.49}{5,000} = 0.442898$$

Now add 1 = 1.442898

Using a calculator with a Y^x key do the following.

1.442898 | inverse or 2nd function key |
| Y^x key | 3 | = | 1.13 subtract 1.00 = 0.13

The rate of interest is 13 %

When you do the inverse of Y^x you are doing the function $\sqrt[x]{y}$, which is the number Y to the root of X.

Continued

Money - Interest

Determining Interest Rate *(continued)*

Example - Interest Compounded Monthly

The original amount was $ 5,000 and no payments had been made for 3 years. After 3 years the total amount due was $ **7,369.43** . What was the rate of interest if the interest was compounded **monthly**?

The amount of interest is

$$7,369.43 - 5,000 = \$ 2,369.43$$

The amount of interest is then divided by the original amount.

$$\frac{2,369.43}{5,000} = 0.473886$$

Now add 1 = 1.473886

Using a calculator with a Y^x key do the following.

1.473866 | inverse or 2nd function key |

| Y^x key | 36 | = | 1.0108333   subtract 1.00 = 0.0108333

in the above 12 months X 3 years = 36

0.0108333 X 12 months = 13

The interest rate was 13 %

Please note that by compounding the interest rate monthly over yearly the amount increased $ 154.94

Continued

Determining Interest Rate *(continued)*

Example - Interest Compounded Daily

The original amount was $ 5,000 and no payments had been made for 3 years. After 3 years the total amount due was $ **7,384.39** . What was the rate of interest if the interest was compounded **daily**?

The amount of interest is

$$7,384.39 - 5,000 = \$ 2,384.39$$

The amount of interest is then divided by the original amount.

$$\frac{2,384.39}{5,000} = 0.4768783$$

Now add 1 = 1.4768783

Using a calculator with a Y^x key do the following.

1.4768783 | inverse or 2nd function key |

| Y^x key | 1095 | = | 1.003562   subtract 1.00 = 0.003562

in the above 365 X 3 years = 1095

0.0003562 X 365 days = 0.13

The interest rate was 13 %

Please note that by compounding the interest rate daily instead of monthly, the amount increased $ 14.96 and daily over yearly it increased $ 169.90.

Money - Interest

Estimating the Interest Payment- There are situations where estimating the interest payment maybe necessary because of the payment structure.

> **Example-** A customer agrees to buy a machine for $5,000 and will pay $1,000.00 a month at 12 % annual interest. To get an accurate tally of the amount of interest due you could do the following table.
>
> $$\frac{\text{annual rate}}{\text{months per year}} \quad \frac{.12}{12} = 0.01 \text{ interest per month}$$
>
> 5,000.00 X 1.01 = 5,050.00
> 5,050.00 - 1,000 = 4,550.00
> 4,050.00 X 1.01 = 4,090.50
> 4,090.50 - 1,000 = 3,090.50
> 3,090.50 X 1.01 = 3,121.41
> 3,121.41 - 1,000 = 2,121.41
> 2,121.41 X 1.01 = 2,142.62
> 2,142.62 - 1,000 = 1,142.62
> 1,142.62 X 1.10 = 1,154.05
> 1,154.05 - 1,000 = 154.05   interest due

The above table is for only 5 payments, had it been for 20 payments the table would be 4 times longer. To save time do the following formula.

Determining Interest Payment on Unstructured Loan

$$\begin{array}{c}\text{Amount} \\ \text{of Interest}\end{array} = \begin{array}{c}\text{Loan} \\ \text{amount}\end{array} \times \left(1 + \frac{\text{rate}}{12}\right)^{\text{number of payments}} - \begin{array}{c}\text{Monthly} \\ \text{Payment}\end{array} \times \frac{\left(1 + \frac{\text{rate}}{12}\right)^{\text{number of payments}} - 1}{\frac{\text{rate}}{12}}$$

(on the next page the above example will be worked)

Determining Interest Payment on Unstructured Loan

Example- $ 5,000 is loaded at 12 % interest and 5 monthly payments of $ 1,000 are to be made. How much will the interest be? *(see example on previous page)*

$$\text{Amount of Interest} = 5{,}000 \times \left(1 + \frac{.12}{12}\right)^{5} - 1{,}000 \times \frac{\left[\left(1 + \frac{.12}{12}\right)^{5}\right] - 1}{\frac{.12}{12}}$$

$$\text{Amount of Interest} = 5{,}000 \times \left(1 + 0.01\right)^{5} - 1{,}000 \times \frac{\left[\left(1 + 0.01\right)^{5}\right] - 1}{0.01}$$

$$\text{Amount of Interest} = 5{,}000 \times \left(1.01\right)^{5} - 1{,}000 \times \frac{\left[\left(1.01\right)^{5}\right] - 1}{0.01}$$

calculate $1.01 \boxed{Y^x \text{ key}}\ 5 \boxed{=} 1.0510101$

$$\text{Amount of Interest} = \left(5{,}000 \times 1.0510101\right) - 1{,}000 \times \frac{1.0510101 - 1}{0.01}$$

$$\text{Amount of Interest} = \left(5{,}000 \times 1.0510101\right) - 1{,}000 \times \frac{0.0510101}{0.01}$$

$$\text{Amount of Interest} = 5{,}255.05 - 1{,}000 \times 5.10101$$

$$\text{Amount of Interest} = 5{,}255.05 - 5{,}101.01$$

$$\text{Amount of Interest} = \$ 154.04 \text{ is due after the last payment is made}$$

Money - Interest

Balance/Interest

Usually a credit card company advertises its annual rate of interest. You can then figure the amount of interest you will pay per month.

APR 15.6 divided by 12 = 1.3 % per month
If your balance is $560.00 X 0.013 = $ 7.28 interest.
You can quickly figure your balance by adding 1 to the above interest rate.

$ 560.00 X 1.013 = $ 567.28

Difference in Interest Rates

Example- if $100.00 is the balance in two different accounts; one account is at 13.9 % and the other is twice as high at 27.8%. How much money will the higher rate cost?

0.139 divided by 12 = .0115833 add 1 = 1.0115833
1.0115833 $\boxed{Y^X \text{ key}}$ 12 $\boxed{=}$ 1.1482065
1.1482065 X 100.00 = $ 114.82
0.278 divided by 12 = 0.0231667 add 1 = 1.0231667
1.0231667 $\boxed{Y^X \text{ key}}$ 12 $\boxed{=}$ 1.3165052
1.3163052 X 100.00 = $ 131.63

The second interest rate was twice as high (200%), as the first however it generated 213% more interest.

$$\frac{\$ \, 31.63}{\$ \, 14.82} = 2.13$$

Penalty Fees-

Example- A person makes a payment 20 days late incurring a $ 30.00 late payment penalty.
The payment could have been as low as $40.00.
What was the annual rate of interest for the use of $40.00 for 20 days?

$$\frac{30}{40} = .75 \qquad \frac{365}{20} = 18 \qquad 18 \text{ X} .75 = 13.50$$
$$13.5 = 1350 \%$$

The interest was 75 %, an annual rate of 1,350 %

Time Required for Investment to Equal Greater Amount

$$\frac{(Future\ value)\ Log\ -\ (Present\ value)\ Log}{(\ 1 + Rate\ )\ Log}$$

This formula requires a calculator that has a "Log" key

Example-You have $ 20,000 invested at 6 % compounded yearly. How long will it take to equal $ 30,000 if no other deposits except interest are made?

$$\frac{(\ 30,000\ )\ Log\ -\ (\ 20,000\ )\ Log}{(\ 1 + .06\ )\ Log}$$

Inter 30,000 into calculator then push Log key = 4.4771213
Then do the same for 20,000 = 4.30103 and 1.06 = .0253059

$$\frac{4.4771213 - 4.30103}{.0253059} = \frac{0.1760913}{.0253059} = 6.95 \quad or\ 7\ years$$

The same problem using Table A without Logarithms

$$Time = \left(\frac{Rate}{\substack{Number\ of\ times \\ compounded\ in\ year}} \right) and \left(\frac{Future\ Value}{Present\ Value} \right)$$

$$Time = \left(\frac{.06}{1} \right) and \left(\frac{30,000}{20,000} \right)$$

$$Time = (\ .06\ ) \quad and \quad (\ 1.5\ )$$

Go to Table A (page 10:1) column 1.06 and go down until you come to a number close to 1.5 . The closest is 1.5036 , now go to the left to the N column to number 7 . The number of compounding periods is 7. In the above example interest was compounded once a year so, Time = 7 years.

Time Required for Investment to Equal Greater Amount without A Log Key, Logarithm Chart or Charts

Example - You have $ 2,000 in the bank compounding interest at 7 % interest. How long will it take to total $ 3,000?

Step 1   Subtract the principal from the amount that you want. This is the amount of interest.

$ 3,000 - 2000 = $ 1000

Step 2   Divide the principal into the interest

$$\frac{1,000}{2,000} = .5 \quad \text{then add } 1 = 1.5$$

Step 3   Divide the interest rate by the number of times interest is compounded per year.

$$\frac{.07}{12} = .005833 \quad \text{then add } 1 = 1.005833$$

Step 4   Using a calculator with a $\boxed{Y^X}$ key do the following.

1.0058333 $\boxed{Y^X}$ any number until it equals 1.5

1.0058333 $\boxed{Y^X}$ *try* 10 $\boxed{=}$ 1.0598883 *which is way to low*

1.0058333 $\boxed{Y^X}$ *try* 120 $\boxed{=}$ 2.0096534 *to high*

1.0058333 $\boxed{Y^X}$ *try* 100 $\boxed{=}$ 1.7889614 *slightly to high*

1.0058333 $\boxed{Y^X}$ *try* 65 $\boxed{=}$ 1.4594547 *close*

1.0058333 $\boxed{Y^X}$ *try* 70 $\boxed{=}$ 1.5025214 *close enough*

(continued)

Time (continued)

Step 5 Divide the power number by the number of times interest is compounded per year.

$$\frac{70}{12} = 5.83 \text{ years}$$

Step 6 We will now check our answer.

$$\$\,2{,}000 \text{ X } \quad 1 + \left(\frac{.07}{12}\right)^{\overset{5.83 \text{ years X 12 compound periods}}{}}$$

5.83 X 12 = 69.96

$$\$\,2{,}000 \text{ X } 1.0058333^{69.96}$$

calculate 1.0058333 Y^x 69.96 = 1.5021719

$$\$\,2{,}000 \text{ X } 1.5021719 = \$\,3004.34$$

Time Needed to Double Your Money

The Quick Way Using the magic number 70
(this number is only useful when doubling your money, or return on investment, and interest is compounded yearly).

Divide 70 by the rate of interest or percent of return on investment.

$$\frac{70}{\text{Rate of interest or Rate of Return}}$$

Example : How long will it take to double your money at 5 % interest, compounded yearly.

$$\frac{70}{.05} \quad \text{Change to} \quad \frac{70}{5} = 14 \text{ years}$$

Money - Time

Time for Monthly Payment to Equal Determined Amount

> **Example-** A life insurance policy pays $ 10,000 upon
> death. The monthly payment is $ 35.50. How long
> would it take, if you invest $ 35.50 monthly at 10 %
> annual interest, compounded monthly, to equal
> $ 10,000?

Step 1 Divide the payout by the payment. $\dfrac{10,000}{35.50}$ = 281.69

Step 2 Determine the monthly interest. $\dfrac{0.10}{12}$ = 0.0083333

Step 3 Multiply Step 2 by Step 3
$$281.69 \times 0.0083333 = 2.3473$$

Step 4 Add 1 to 2.3473 = 3.3473

Step 5 Add 1 to the monthly rate of 0.0083333 = 1.0083333

Step 6 Guess/estimate how many months it would take. To
start with we will use 120 months (ten years).

Step 7 Enter the following into a calculator. 1.00833
$\boxed{Y^X \text{ key}}$ 120 $\boxed{=\text{key}}$ 2.7069

Step 8 Keep entering a number after the Y^X key until you
have an amount close to Step 4. (3.3473)
$$1.008333 \ Y^X \ \ 146 = 3.35877 \text{ is close enough}$$

Step 9 Divide 146 by 12 months = 12.1667 = 12 yrs 2 months

> It takes 12 years and 2 months of investing $ 35.50 per
> month, at 10 % interest compounded monthly to equal
> $ 10,000.

Savings Account with Compounded Interest
Only One Deposit Made at Beginning

Amount of 1 at Compound Interest

Compounding interest means paying interest, adding it to the principal and then paying interest on the new principal, that includes the original principal, but now also includes the interest just earned.

Compounding (paying and crediting) interest adds numerous steps to the simple formula of

Savings = Principal X Rate X Time

Example - How much will you have in the bank if you put in $2,000 for 1 year and the bank compounds the interest twice during the year, at 10 % annually?

Step 1 Divide the annual interest rate by the number of times it is compounded during the year.

$$\frac{.10}{2} = .05$$

Step 2 Figure the amount of interest paid for the first of the interest pay periods.

2,000 X .05 = 100

Step 3 Add that to the principal

2,000 + 100 = 2,100

Step 4 Figure the amount of interest paid for the next interest pay period.

2,100 X .05 = 105

Step 5 Add that to the principal

2100 + 105 = $2205 is the amount of money you will have after 1 year.

As in the previous example we figured the interest and new principal for each of the compounded periods.

Another way of doing it is to add 1 to the rate and

1.05 X 1.05 = 1.1025

Then multiplying this with the principal

$ 2,000 X 1.1025 = $ 2,205

Money - Savings

Amount of 1 At compound Interest (continued)

Another way is to look up the compounded amount on an Amount of 1 at Compound Interest table. A short version of this table is in Chapter 10 :1-5 as Table A. At the top find column 1.05, then go down column N to 2. Go across and they intersect at 1.1025

Using a calculator with a $\boxed{Y^X}$ or $\boxed{X^Y}$ key we can also find the compounded amount. This key will raise the interest rate to the x or y power.
On a calculator do the following

1.05 $\boxed{Y^X}$ 2 $\boxed{=}$ 1.1025 appears

Now we will do all three methods again, using 24 % annual interest, for two years compounded 4 times a year.

$$\frac{.24}{4 \text{ times}} = 0.06$$

First Method

1.06 X 1.06 X 1.06 X 1.06 X 1.06 X 1.06 X 1.06 X 1.06 = 1.593848

Second Method using Table A

Using Table A go to column 1.06 and column N and go down to 8 (2 years X 4 periods) intersecting at 1.593848

Third Method using a calculator with a $\boxed{Y^X}$ key.

1.06 $\boxed{Y^X}$ 8 $\boxed{=}$ 1.593848

If we were to deposit $2,000 we would multiply it by 1.593848 equaling $3187.7

Now we will put the complete formula to work in the next page.

Savings (continued)

Number of Times Compounded
During the entire savings program

$$\text{Savings} = \text{Principal X} \left(1 + \frac{\text{Rate}}{\text{Number of Times Compounded During a year}} \right)$$

Example - $2,000 is deposited for one year and it is compounded 2 times a year at 12 % interest.

Using Table A

$$\text{Savings} = 2,000 \text{ X } \left(1 + \frac{.12}{2} \right)^{2 \text{ times}} \qquad \textit{.12 divided by 2 = .06}$$

$$\text{Savings} = 2,000 \text{ X } \left(1 + .06 \right)^{2 \text{ times}}$$

See Table A page 10 :1   Fine the 1.06 column and go down to N2 and fine the number 1.1236
Savings = 2,000 X 1.1236
Savings = $ 2,247.20

Using a Calculator with a $\boxed{Y^X}$ key.

1.06 $\boxed{Y^X}$ 2 $\boxed{=}$ 1.1236

2,000 X 1.1236 = $ 2247.2

Example - The same problem but this time we will figure it with the bank compounding interest 12 times a year for 3 years a 7% interest.
Savings = ????
Principal = $2,000
Rate = .07
Time = 3 years
Compounded monthly = 12 times

$$\text{Savings} = 2,000 \text{ X } \left(1 + \frac{.07}{12} \right)^{3 \text{ years X 12 times}}$$

$$\text{Savings} = 2,000 \text{ X } \left(1.0058333 \right)^{36} \qquad \textit{(continued)}$$

Savings *(continued)*

Find 1.005833 in Table B page 10:6 and go down the N
column to 36. Go across to 1.2329241
   Savings = 2,000 X 1.2329241 = 2,465.85
   Savings = $2,465.85

Using a Calculator with a $\boxed{Y^X}$ key.

$$2{,}000 \text{ X}\left(1 + \frac{.07}{12}\right)^{36}$$

$$2{,}000 \text{ X }(1 + .0058333)^{36}$$

Calculate 1.0058333 $\boxed{Y^X}$ 36 $\boxed{=}$ 1.2329241

 2,000 X 1.2329241 = 2,465.85

Example - $2,000 deposited in the bank at .05 annual
  interest for 10 years interest compounded
  monthly (12 times a year).

$$\text{Savings} = 2{,}000 \text{ X}\left(1 + \frac{.05}{12}\right)^{12 \text{ times a year X 10 years}}$$

Savings = 2,000 X (1 .004167)^{120}
Go to Table B column 1.0041667 and column N 120
intersecting at 1.6470161
Savings = 2,000 X 1.6470161 = 3,294.03
Savings = $ 3,294.03

Using a calculator with a $\boxed{Y^X}$ key.

$$\frac{.05}{12} = .0041667 + 1 = 1.0041667$$

1.0041667 $\boxed{Y^X}$ 120 $\boxed{=}$ 1.6470161

1.6470161 X 2,000 = $ 3,294.03

Compounding Interest Daily Versus Monthly or Yearly

Interest Compounded <u>Daily</u>

Example - $ 10,000 compounded daily for 1 year at 6 % interest. How much would you have after 1 year?. (Use Table C page 10:10-12)

$$10,000 \times \left(1 + \frac{.06}{365} \right)^{365}$$ *.06 divided by 365 = .000164*
then add the 1

$$10,000 \times 1.0001644^{365}$$

$$1.0001644 \boxed{Y^X} 365 \boxed{=} 1.0618313$$

$$10,000 \times 1.0618313 = \$ 10,618.31$$

Interest Compounded <u>Monthly</u>

$$10.000 \times \left(1 + \frac{.06}{12} \right)^{12}$$ *use Table B*

.06 divided by 12 = .005
then add the 1

$$10,000 \times 1.005^{12}$$

$$1.005 \boxed{Y^X} 12 \boxed{=} 1.0616778$$

$$10,000 \times 1.0616778 = \$ 10,616.78$$

Interest Compounded <u>Yearly</u>

$$10,000 \times 1.06 = \$ 10,600$$

Results

Compounded Daily- $10,618.31. $18.31 more than annually
Compounded Monthly - $10,616.78. $16.78 more than annually
Compounded Yearly -   10,600.00

You yield 3.05% more with compounding daily and 2.78% more compounding monthly over compounding annually.

Saving Monthly for a Future Amount

To determine how much has to be deposited monthly for a specific amount by a specific date.

Step 1

$$\text{Amount to Deposit Per Month} = \frac{\left(1 + \dfrac{\text{Rate}}{\substack{\text{Compound periods} \\ \text{in a year}}}\right)^{\text{Number of deposits} + 1} - 1}{\dfrac{\text{Rate}}{\text{Compound periods in a year}}} - 1$$

This formula is similiar to the one on page 9:33

Step 2 The answer to the above is then divided into the amount needed.

Example - $ 1,000 will be needed in 1 year, how much should be deposited monthly at 8 % interest compounded monthly.

$$\text{Amount to Deposit Per Month} = \left\{ \frac{\left[\left(1 + \dfrac{.08}{12}\right)^{12 + 1}\right] - 1}{\dfrac{.08}{12}} \right\} - 1$$

$$\text{Amount to Deposit Per Month} = \left\{ \frac{\left[\left(1.0066667\right)^{13}\right] - 1}{.0066667} \right\} - 1$$

$$1.0066667 \ \boxed{Y^x} \ 13 \ \boxed{=} \ 1.090219504$$

$$\text{Amount to Deposit Per Month} = \left\{ \frac{\left[1.090219504\right] - 1}{.0066667} \right\} - 1$$

$$\text{Amount to Deposit Per Month} = \left\{ \frac{.090219504}{.0066667} \right\} - 1$$

$$\text{Amount to Deposit Per Month} = 13.53285787 \quad - 1 \quad \textit{(continued)}$$

Saving Monthly for a Future Amount (*continued*)

Step 2

$$\frac{\text{Amount to Deposit}}{\text{Per Month}} = \frac{\text{Amount needed}}{12.53285787}$$

$$\frac{\text{Amount to Deposit}}{\text{Per Month}} = \frac{\$\,1,000}{12.53285787}$$

1,000 divided by 12.53285787 equals $ 79.79

Another Example - $ 10,000 will be needed in 7 years
How much should be deposited annually
at 7 % compounded annually

This formula is similar to page 9:29. Instead of multiplying the amount deposited you divide into the amount needed.

Step 1

$$\frac{\text{Amount to Deposit}}{\text{Per Month}} = \left\{ \frac{\left[\left(1 + \frac{.07}{1}\right)^{7+1}\right] - 1}{\frac{.07}{12}} \right\} - 1$$

$$\frac{\text{Amount to Deposit}}{\text{Per Month}} = \left\{ \frac{\left[\left(1.07\right)^{8}\right] - 1}{.07} \right\} - 1$$

calculate $1.07 \boxed{Y^X} \; 8 \; \boxed{=} \; 1.71818618$

$$\frac{\text{Amount to Deposit}}{\text{Per Month}} = \left\{ \frac{1.71818618 - 1}{.07} \right\} - 1$$

$$\frac{\text{Amount to Deposit}}{\text{Per Month}} = \left\{ \frac{.71818618}{.07} \right\} - 1$$

$$\frac{\text{Amount to Deposit}}{\text{Per Month}} = 10.25980257 - 1$$

Step 2 $ 10,000 divided by 9.25980257
equals $ 1,079.94

Money - Savings

Savings / Annuity Programs
Deposits Made Periodically with Compound Interest

Annuities or savings programs, depend upon regular scheduled payments or deposits over a period of time. There are two annuities that require almost the same formula but are slightly different.

Ordinary Annuity- Payment or deposit is made at the end of each period. For example, if the annuity is to run 5 years the payment would be at the end of each year. There would be only four interest earning periods. If deposits are monthly the payments would be at the end of each month. For example, if the program was five years with deposits made monthly, there would be 59 interest earning periods (5 X 12 = 60 - 1 = 59).

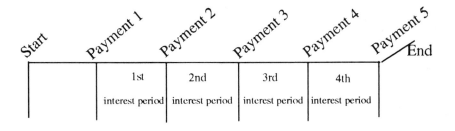

The above illustrates there are 4 interest paying periods when the payment is made at the end of each period.

Annuity Due - A Regular Type Savings Program
Payment or deposit is made at the beginning of each period. If an Annuity Due runs for 5 years, with deposits made yearly, there would be 5 interest earning periods. Likewise for 5 years and deposits made monthly, there would be 60 interest earning periods. This assumes that interest is

(continued)

Annuity Due (continued)

to be paid and compounded the same as the deposit periods. For example, an Annuity Due of five years would be deposited at the beginning of each year equaling 5 deposits and there would be 5 interest earning periods.

1st	2nd	3rd	4th	5th
interest period	interest period	interest period	interest period	interest period

The above illustrates 5 payment periods and 5 interest earning periods . If interest was compounded monthly there would be 60 interest compounding periods.

When the actual formulas are done, you would think that you would have to compensate for the Ordinary Annuity, because it has less interest earning periods than the Annuity Due does. However, it is just the opposite for how the formulas workout. You have to add an extra compounding period and then subtract a number one at the final math operation, when you do the formula for Annuity Due.

Formulas for Finding the Future Value of Savings Programs (Annuities - FVA)

The following pages have these formulas

Ordinary Annuity Deposits Annually - Compounded Annually
Annuity Due   Deposits Annually - Compounded Annually
Ordinary Annuity Deposits Monthly - Compounded Monthly
Annuity Due   Deposits Monthly - Compounded Monthly
Ordinary Annuity Deposits Annually - Compounded Monthly
Annuity Due   Deposits Annually - Compounded Monthly

Money - Savings

Ordinary Annuity - Deposited Annually
Compounded Annually

The Math Formula is on the following page.

The Deposit is made one year <u>after</u> the program has started.
The program ends when the last payment is made.

In the example the program starts on January 1st, however, the first deposit or payment is not until the end of the year, on December 31st. $ 300 is deposited annually at 7% interest for 5 years.

The interest rate of 7% is changed to .07 and 1 is added to it equaling 1.07. 1.07 is multiplied times it's self 5 times, the number of payments made.

On a calculator do the following.

Enter 1.07 $\boxed{Y^x}$ 5 $\boxed{=}$ 1.4025517

Or go to Table A   Find 1.07,   go down Column N to 5 and across to 1.4025517

1 is subtracted from 1.4025517 equaling .4025517

.4025517 is divided by the interest rate .07 equaling 5.750739

5.750739 is then multiplied by $300 equaling $ 1,725.22

Money - Savings

Ordinary Annuity - Deposited Annually
Compounded Annually

*(the instructions are
on the previous page)*

The first deposit is made 1 year after the program starts.

$$FVA = \text{Amount Deposited} \times \frac{(1 + \text{Rate})^{\text{Number of Deposits}} - 1}{\text{Rate}}$$

Example - $ 300 Deposited annually, 7 %
interest compounded annually
How much will you have after 5
years?

$$FVA = \$ 300 \times \frac{\left[(1 + .07)^5\right] - 1}{.07}$$

Calculate $1.07 \boxed{Y^X} \; 5 \; \boxed{=} \; 1.4025517$

$$FVA = 300 \times \frac{1.4025517 - 1}{.07}$$

$$FVA = 300 \times \frac{.4025517}{.07}$$

$$FVA = 300 \times 5.750739$$

$$FVA = \$ 1,725.22$$

Money - Savings

Typical Savings Program
Annuity Due - Deposited Annually
Compounded Annually

The Math Formula is on the following page

The first deposit is made on the first day the program is started.
The program ends one year after the last payment is made.

In the example, the program starts on January 1st when the
first deposit or payment is made and ends 5 years later on
December 31st. $ 300 is deposited annually at 7% interest for
5 years.

The interest rate of 7% is changed to .07 and 1 is added to it
equaling 1.07. 1.07 is multiplied times it's self 6 times, (the
number of payments made plus one extra compounding period
equaling 6).

On a calculator do the following.

Enter 1.07 $\boxed{Y^x}$ 6 $\boxed{=}$ 1.5007304

Or go to Table A (page 10:1) - Find 1.07, go down Column N
to 6 and across to 1.5007304

1 is subtracted from 1.5007304 equaling .5007304

.5007304 is divided by the interest rate .07 equaling 7.1532907

1 is subtracted from 7.1532907 equaling 6.1532907

6.1532907 is then multiplied by $300 equaling $ 1,845.99

Typical Savings Program
Annuity Due - Deposited Annually
Compounded Annually

(the instructions are on the previous page)

Example - $ 300 Deposited annually, 7 % interest compounded annually. How much will you have after 5 years?

$$FVA = \frac{Amount}{Deposited} \times \frac{(1 + Rate)^{Number\ of\ deposits\ +\ 1} - 1}{Rate} - 1$$

(compound period)

$$FVA = 300 \times \left\{ \left[\frac{(1 + .07)^{5+1}}{.07} \right] - 1 \right\} - 1$$

$$FVA = 300 \times \left\{ \frac{[(1.07)^6] - 1}{.07} \right\} - 1$$

Calculate $1.07 \boxed{Y^X} 6 \boxed{=} 1.5007304$

$$FVA = 300 \times \left\{ \frac{1.5007304 - 1}{.07} \right\} - 1$$

$$FVA = 300 \times \left\{ \frac{.5007304}{.07} \right\} - 1$$

$$FVA = 300 \times 7.1532907 - 1$$

$$FVA = 300 \times 6.1532907$$

$$FVA = \$ 1,845.99$$

Money - Savings

Ordinary Annuity - Deposited Monthly
Compounded Monthly

The Math Formula is on the following page.

The Deposit is made one month <u>after</u> the program has started. The program ends when the last payment is made.

In the example, the program starts on January 1st, however, the first deposit or payment is not made until January 31. The program ends 5 years later on December 31st, when the last payment is made. $ 2,000 is deposited monthly at 7% interest for 5 years.

The interest rate of 7% is changed to .07 and is divided by the number of compounding periods in a year, .07 divided by 12 = .0058333. This is done twice, once for the top part of the equation and once for the lower part. For the above part 1 is added to it equaling 1.0058333. This number is then multiplied times it's self the number of payments made, which is 60 equaling 1.4176224

On a calculator do the following.

Enter 1.0058333 $\boxed{Y^x}$ 60 $\boxed{=}$ 1.4176224

Or go to Table B (page 10:6) Find 1.0058333 go down Column N to 60 and across to 1.4176224.

1 is subtracted from 1.4176224 equaling .4176224

.4176224 is divided by .0058333 equaling 71.592827

71.592827 is then multiplied by $ 2,000 equaling $ 143,185.66

Ordinary Annuity - Deposited Monthly
Compounded Monthly

$$\text{Amount Deposited} \times \dfrac{\left(1 + \dfrac{\text{Rate}}{\text{Compound periods in a year}}\right)^{\text{Number of Deposits}} - 1}{\dfrac{\text{Rate}}{\text{Compound periods in a year}}}$$

Example - $ 2,000 Deposited Monthly, with 7 % interest compounded monthly. How much will you have after 5 years?

$$FVA = \$\, 2,000 \times \dfrac{\left[\left(1 + \dfrac{.07}{12}\right)^{(12 \times 5)}\right] - 1}{\dfrac{.07}{12}}$$

$$FVA = 2,000 \times \dfrac{\left[1.0058333^{60}\right] - 1}{.0058333}$$

Calculate $1.0058333 \boxed{Y^x}\ 60 \boxed{=}\ 1.4176224$

$$FVA = 2,000 \times \dfrac{\left[1.4176224\right] - 1}{.0058333}$$

$$FVA = 2,.000 \times \dfrac{.4176224}{.0058333}$$

$$FVA = 2,000 \times 71.592827$$

$$FVA = 143,185.65$$

Money - Savings

Typical Savings Program or
Annuity Due - Deposited Monthly
Compounded Monthly

The Math Formula on the following page.

The First deposit is made the day the program started.
The program ends one month after the last payment is made.

In the example the program starts on January 1st when the first
deposit or payment is made. The program ends 5 years later
on December 31st. $ 2,000 is deposited monthly at 7%
interest for 5 years.

The interest rate of 7% is changed to .07 and is divided by the
number of compounding periods in a year .07 divided by 12 =
.0058333. This is done twice, once for the top part of the
equation and once for the lower part. For the above part 1 is
added to it equaling 1.0058333. This number is then multiplied
times it's self the number of compounding periods (60) plus 1
for an additional compounding period equaling 1.4258919

On a calculator do the following.

Enter 1.0058333 $\boxed{Y^x}$ 61 $\boxed{=}$ 1.4258919

Or go to Table B (page 10:6) Find 1.0058333 go down
Column N to 61 and across to 1.4258919.

1 is subtracted from 1.4258919 equaling .4258919

.4258919 is divided by .0058333 equaling 73.01046

1 is subtracted from 73.01046 equaling 72.01046

72.01046 is then multiplied by $ 2,000 equaling $ 144,020.92

Typical Savings Program
Annuity Due - Deposited Monthly
Compounded Monthly

$$\frac{\text{Amount}}{\text{Deposited}} \times \frac{\left(1 + \dfrac{\text{Rate}}{\text{Compound periods in a year}}\right)^{\substack{\text{Number of Deposits} \\ + \ 1 \ \text{month}}} - 1}{\dfrac{\text{Rate}}{\text{Compound period in a year}}} - 1$$

Example- $ 2,000 Deposited Monthly, 7 % interest compounded monthly. How much will you have after 5 years?

$$FVA = \$ 2,000 \times \left\{ \frac{\left[\left(1 + \dfrac{.07}{12}\right)^{12 \, X \, 5 \, + \, 1}\right] - 1}{\dfrac{.07}{12}} \right\} - 1$$

$$FVA = 2,000 \times \left\{ \frac{\left[1.0058333^{\,61}\right] - 1}{.0058333} \right\} - 1$$

Calculate 1.0058333 $\boxed{Y^X}$ 61 $\boxed{=}$ 1.425819

$$FVA = 2,000 \times \left\{ \frac{\left[1.4258919 - 1\right]}{.0058333} \right\} - 1$$

$$FVA = 2,000 \times \left\{ \frac{.4258919}{.0058333} \right\} - 1$$

$$FVA = 2,000 \times \ 73.01046 \ - 1$$

$$FVA = 2,000 \times \ 72.01046$$

$$FVA = \$ \ 144,020.92$$

Money - Savings

Ordinary Annuity - Deposited Annually
Compounded Monthly

The Math Formula is on the following page.

The Deposit is made one year <u>after</u> the program has started. The program ends when the last payment is made. Even though the interest is compounded monthly, keep in mind that during the first year there is nothing deposited to draw interest.

In the example the program starts on January 1st, however, the first deposit or payment is not made until December 31. The program ends 5 years later on December 31st, when the last payment is made. $ 2,000 is deposited monthly at 7% interest for 10 years.

The interest rate of 7% is changed to .07 and is divided by the number of compounding periods in a year, .07 divided by 12 = .0058333. This is done twice, once for the top part of the equation and once for the lower part. For each part, 1 is added to it equaling 1.0058333. For the upper part 1.0058333 will be multiplied times itself the number of years, 10 and the number of compounding periods in a year, 12 equaling 120. 1.0058333 multiplied times itself 120 time equals 2.0096534

On a calculator do the following.
Enter 1.0058333 $\boxed{Y^x}$ 120 $\boxed{=}$ 2.0096534

Or go to Table B (page 10:6) Find 1.0058333 go down Column N to 120 and across to 2.0096534

The lower 1.0058333 is times itself the number of compounding periods in one year. 1.0058333 times itself 12 times equals 1.0722897

1 is subtracted from 2.0096534 equaling 1.0096534
1 is subtracted from 1.0722897 equaling .0722897
1.0096534 is divided by .0722897 equaling 13.966767

13.966767 is then multiplied by $ 2,000 equaling $ 27,933.53

Ordinary Annuity - Deposited Annually
Compounded Monthly

$$FVA = \text{Amount Deposited} \times \dfrac{\left(1 + \dfrac{\text{Rate}}{\text{Compound periods in a year}}\right)^{\text{Compound periods during program}} - 1}{\left(1 + \dfrac{\text{Rate}}{\text{Compound periods in a year}}\right)^{\text{Compound periods in a year}} - 1}$$

Example - $ 2,000 Deposited annually with 7 % interest compounded monthly. How much will you have after 10 years?

$$FVA = 2,000 \times \left\{ \dfrac{\left[\left(1 + \dfrac{.07}{12}\right)^{10 \times 12}\right] - 1}{\left[\left(1 + \dfrac{.07}{12}\right)^{12}\right] - 1} \right\}$$

$$FVA = 2,000 \times \left\{ \dfrac{\left[(1.005833)^{120}\right] - 1}{\left[(1.0058333)^{12}\right] - 1} \right\}$$

$$1.0058333 \boxed{Y^X} 120 = 2.0096534$$

$$1.0058333 \boxed{Y^X} 12 = 1.0722897$$

$$FVA = 2,000 \times \left\{ \dfrac{2.0096534 - 1}{1.0722897 - 1} \right\}$$

$$FVA = 2,000 \times \left\{ \dfrac{1.0096534}{.0722897} \right\}$$

$$FVA = 2,000 \times 13.966767 = \$ 27,933.53$$

Money - Savings

Typical Savings Program
Annuity Due - Deposited Annually
Compounded Monthly

The Math Formula is on the following page.

The Deposit is made the first day the program has started.
The program ends one year after last payment is made.

In the example, the program starts on January 1st, when the
first deposit or payment is made, until December 31 ten years
later. $ 2,000 is deposited monthly at 7% interest for 10 years.
The interest rate of 7% is changed to .07 and is divided by the
number of compounding periods in a year, .07 divided by 12 =
.0058333. This is done twice, once for the top part of the
equation and once for the lower part. For each part, 1 is added
to it equaling 1.0058333. For the upper part 1.0058333 will be
multiplied times itself, the number of years (10) and the
number of compounding periods in a year (12) equaling 120.
There are 12 additional compounding periods added to 120
equaling 132. 1.0058333 multiplied times itself 132 times
equals 2.1549305

On a calculator do the following or use Table B

Enter 1.0058333 $\boxed{Y^x}$ 132 $\boxed{=}$ 2.1549305

The lower 1.0058333 is times itself the number of
compounding periods in one year. 1.0058333 times itself 12
times equals 1.0722897

1 is subtracted from 2.1549305 equaling 1.1549305
1 is subtracted from 1.0722897 equaling .0722897
1.1549305 is divided by .0722897 equaling 15.976418
1 is subtracted from 15.976418 equaling 14.976418

14 .976418 is then multiplied by $ 2,000 equaling $ 29,952.84

Typical Savings Program
Annuity Due - Deposited Annually
Compounded Monthly

$$FVA = \frac{Amount}{Deposited} \times \frac{\left(1 + \dfrac{Rate}{Compound\ periods\ in\ a\ year}\right)^{\substack{Compound\ periods \\ during\ program \\ +\ 12\ Compound\ periods}} - 1}{\left(1 + \dfrac{Rate}{Compound\ periods\ in\ a\ year}\right) - 1} - 1$$

Example - $ 2,000 Deposited annually with 7 %
interest compounded monthly. How much
will you have after 10 years?

$$FVA = \$\,2{,}000 \times \left\{ \frac{\left[\left(1 + \dfrac{.07}{12}\right)^{(10\,X\,12)\,+\,12}\right] - 1}{\left[\left(1 + \dfrac{.07}{12}\right)^{12}\right] - 1} \right\} - 1$$

$$FVA = \$\,2{,}000 \times \left\{ \frac{\left[(1.0058333)^{132}\right] - 1}{\left[(1.0058333)^{12}\right] - 1} \right\} - 1$$

Calculate $1.0058333 \quad \boxed{Y^X} \quad 132 = 2.1549305$

Calculate $1.0058333 \quad \boxed{Y^X} \quad 12 = 1.0722897$

$$FVA = \$\,2{,}000 \times \left\{ \frac{2.1549305 - 1}{1.0722897 - 1} \right\} - 1$$

$$FVA = \$\,2{,}000 \times \left\{ \frac{1.1549305}{.0722897} \right\} - 1$$

$$FVA = \$\,2{,}000 \times 15.976418 - 1$$

$$FVA = \$\,2{,}000 \times 14.976418 = \$\,29{,}952.84$$

Money - Funding

Funding for a Time Limited Period - College or Retirement

When setting up a fund to retire on or send someone to college, you should realize that withdrawals will be made, the fund will draw interest and at the end of the need for the fund, it will be completely depleted.

Example - Your child will need $ 45,000 a year for 4 years to go to college. You have selected an investment that will pay 8.5 % interest compounded annually. How much should you put into the college fund so that upon graduation it is completely depleted? Your child will withdraw $ 45,000 once a year.

$$\text{Amount of Fund} = \text{Annual Withdrawal} \times \frac{1 - \left[(\frac{1}{1 + \text{Rate}})^{\text{number of years}} \right]}{\text{Rate}}$$

$$\text{Amount of Fund} = 45,000 \times \frac{1 - \left[\frac{1}{(1 + .085)^4} \right]}{.085}$$

$$\text{Amount of Fund} = 45,000 \times \frac{1 - \left[\frac{1}{(1.085)^4} \right]}{.085}$$

calculate $1.085 \boxed{Y^x} 4 \boxed{=} 1.3858587$

$$\text{Amount of Fund} = 45,000 \times \frac{1 - \left[\frac{1}{1.3858587} \right]}{.085}$$

1 divided by 1.385857 = .7215742

$$\text{Amount of Fund} = 45,000 \times \frac{\left[1 - .7215742 \right]}{.085}$$

1 minus .7215742 = .2784258

$$\text{Amount of Fund} = 45,000 \times \frac{.2784258}{.085}$$

.2784257 divided by .085 = 3.2755965

$$\text{Amount of Fund} = 45,000 \times 3.2755965 = \$ 147,401.84$$

(continued on next page)

Funding (continued)

Please note that, in the previous example, the right side of the equation is constant, assuming interest rate and time stays the same. If the yearly draw is $30,000 then multiply it times 3.2755965 = $ 98,267.90

Example- When you retire at age 65 you want available $ 40,000 a year for 25 years. This assumes you will live to be 90 at which time the fund will be at $0. You have selected an investment plan that pays 8 % per year compounded yearly. How much do you have to have in the fund 1 year before you retire at age 65?

$$\text{Amount of Fund} = \$40,000 \times \frac{1 - \left[\frac{1}{(1 + .08)^{25}}\right]}{.08}$$

$$\text{Amount of Fund} = \$40,000 \times \frac{1 - \left[\frac{1}{(1.08)^{25}}\right]}{.08}$$

1.08 $\boxed{Y^x}$ *25* $\boxed{=}$ *6.8484752*

$$\text{Amount of Fund} = \$40,000 \times \frac{1 - \left[\frac{1}{6.8484752}\right]}{.08}$$

1 divided by 6.8484752 = .1460179

$$\text{Amount of Fund} = \$40,000 \times \frac{\left[1 - .1460179\right]}{.08}$$

$$\text{Amount of Fund} = \$40,000 \times \frac{.8539821}{.08}$$

1 - .1460179 = .8539821
.8539821 divided by .08 = 10.674776

$$\text{Amount of Fund} = \$40,000 \times 10.674776 = \$426,991.05$$

(continued)

Funding (continued)

      If you live to reach 90 you will have drawn $1,000,000 from the fund that had an original deposit of $ 426,991.05 .

Funding One Year

      In the previous example you received $ 40,000 at the beginning of every year. Naturally, you are not going to spend it all at the beginning of each year. You will probably put it in a savings account and draw on it once a month.

      Example: How much can you withdrawal every month for 12 months when the savings account pays interest of 6 % compounded monthly? Keep in mind after 12 months the account must be completely depleted.

$$\text{Monthly Withdrawal} = \frac{1 - \left[\dfrac{1}{\left(1 + \dfrac{\text{Rate}}{12}\right)^{12}}\right]}{\dfrac{\text{Rate}}{12}} \quad \text{Divided into the amount funded}$$

$$\text{Monthly Withdrawal} = \frac{1 - \left[\dfrac{1}{\left(1 + \dfrac{.06}{12}\right)^{12}}\right]}{\dfrac{.06}{12}} \quad \text{Divided into \$ 40,000}$$

.06 divided by 12 = .005 then add the 1 to make 1.005

$$\text{Monthly Withdrawal} = \frac{1 - \left[\dfrac{1}{\left(1.005\right)^{12}}\right]}{.005} \quad \text{Divided into \$ 40,000}$$

calculate $1.005 \boxed{Y^x} 12 \boxed{=} 1.0616778$

$$\text{Monthly Withdrawal} = \frac{1 - \left[\dfrac{1}{1.0616778}\right]}{.005} \quad \text{Divided into \$ 40,000}$$

continued

Funding One Year (continued)

$$\frac{\text{Monthly}}{\text{Withdrawal}} = \frac{1 - .9419054}{.005} \Bigg\rangle \quad \begin{array}{l}\text{Divided into}\\ \$\,40,000\end{array}$$

$$\frac{\text{Monthly}}{\text{Withdrawal}} = \frac{.0580946}{.005} = 11.61892 \quad \begin{array}{l}\text{Divided into}\\ \$\,40,000\end{array}$$

$$\frac{\text{Monthly}}{\text{Withdrawal}} = \frac{\$\,40,000}{11.61892} = \$\,3442.66 \text{ monthly draw}$$

Funding for Retirement

Example- At the time of retirement, at age 65 you have $521,533. How much can you withdraw monthly, if you have the money invested a 11% compounded monthly for the next 25 years? At the end of 25 years, at age 90, the fund will have $0 in it.

$$\frac{\text{Monthly}}{\text{Withdrawal}} = \frac{1 - \left[\dfrac{1}{\left(1 + \dfrac{\text{Rate}}{12}\right)^{\text{Number of withdraw periods}}}\right]}{\dfrac{\text{Rate}}{12}} \quad \begin{array}{l}\text{Divided into the}\\ \text{amount funded}\end{array}$$

$$\frac{\text{Monthly}}{\text{Withdrawal}} = \frac{1 - \left[\dfrac{1}{\left(1 + \dfrac{.11}{12}\right)^{12 \times 25}}\right]}{\dfrac{.11}{12}} \quad \begin{array}{l}\text{Divided into}\\ \$521,533\end{array}$$

(continued on next page)

Money- Funding

Funding for Retirement (continued)

.11 divided by 12 = .00916667
Then 1 was added to make
1.00916667

$$\text{Monthly Withdrawal} = \frac{1 - \left[\left(\dfrac{1}{1.00916667}\right)^{300}\right]}{.00916667}$$

calculate 1.00916667 $\boxed{Y^X}$ *300* $\boxed{=}$ *15.44788861*

$$\text{Monthly Withdrawal} = \frac{1 - \left[\dfrac{1}{15.44788861}\right]}{.00916667}$$

1 divided by 15.4478861 = .064733776

$$\text{Monthly Withdrawal} = \frac{\left[1 - .064733776\right]}{.00916667}$$

1 -.064733776 = .935266224

$$\text{Monthly Withdrawal} = \frac{.935266224}{.00916667}$$

.935266224 divided by .00916667 = 102.0290389

$$\text{Monthly Withdrawal} = 102.0290389 \quad \text{divide into the amount funded}$$

$$\text{Monthly Withdrawal} = 102.0290389 \text{ divided into } \$521,533$$

$$\text{Monthly Withdrawal} = \$ 5,111.60$$

Funding for Multiple Draws - *without inflation factor*
When a fund is setup, to be drawn from by more than one, at a future date.

Example - Grandparents want to leave $ 100,000 to their 4 grandchildren. Each grandchild is to receive the same amount of money from the fund when they reach the age of 18. The fund will be invested at 8% interest, compounded annually. When the fund is setup their grandchildren are; 16, 14, 11, and 9.For this problem we will consider all children have the same birthday and is was the day before the fund was set up. How much will each be given at age 18?

Determine how many years before each child receives their inheritance at age 18.

1st 18 - 16 = 2, 2nd 18 - 14 = 4, 3rd 18 -11 = 7, 4th 18 - 9 = 9

Determine the present value of each at compound interest.

$$(1.08)^{-2} + (1.08)^{-4} + (1.08)^{-7} + (1.08)^{-9}$$

Using a calculator with $\boxed{Y^X}$ and $\boxed{^+_-}$ key, do the following.

1.08 $\boxed{Y^X}$ 2 $\boxed{^+_-}$ $\boxed{=}$ .8573388

1.08 $\boxed{Y^X}$ 4 $\boxed{^+_-}$ $\boxed{=}$ .7350299

1.08 $\boxed{Y^X}$ 7 $\boxed{^+_-}$ $\boxed{=}$ .5834904   *Add these up*

1.08 $\boxed{Y^X}$ 9 $\boxed{^+_-}$ $\boxed{=}$ .500249

2.6761081

Divide the funded amount by the total.

$$\frac{\$ 100,000}{2.6761081} = \$ 37,367.70$$

Each child will receive $ 37,367.00 when they turn 18.

(continued)

Money - Funding

Funding for Multiple Draws -*without inflation factor*

Check the answer to the pervious example by doing the following-

Fund started with	100,000
After 2 yrs. before 1st child is paid the fund is	
then worth 100,000 X 1.08²	116,640.00
The first child is paid	37,367.70
Balance	79,272.30
After 2 more yrs. the fund is worth 79,272.3 X 1.08²	92,463.21
The second child is paid	37,367.70
Balance	55,095.51
After 3 more yrs. the fund is worth 55,095.51 X 1.08³	69,404.48
The third child is paid	37,367.70
Balance	32,036.78
After 2 more yrs. the fund is worth 32,036.78 X 1.08²	37,367.70
The fourth child is paid	37.367.70
Balance	0

Funding for Multiple Draws - <u>with</u> inflation factored in

Using the previous example we will now factor in an annual inflation rate of 3 %. This will then assure each child (draw), equal buying power.

Take the annual interest rate the fund pays and divide it by the rate of future inflation (estimated). You will have to add a 1 to each of these numbers.

$$\frac{\text{interest rate}}{\text{inflation rate}} \quad \frac{1.08}{1.03} = 1.0485437$$

Now do the same as in the previous example but use this new number in place of the interest rate.

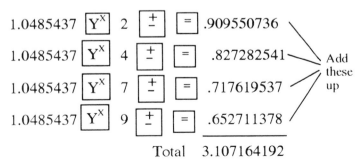

Total   3.107164192

Funding with Multiple Draws - with inflation factored in
(continued)

Divide this new number into the amount funded.

$$\frac{\$\ 100,000}{3.107164192} = 32,183.69$$

Use the amount 32,183.69 to determine the future amount each will draw using the estimated inflation rate.

Each draw will receive

1st draw	1.03	Y^X	2	X	32,183.69	=	$ 34,143.68
2nd draw	1.03	Y^X	4	X	32,183.69	=	$ 36,223.03
3rd draw	1.03	Y^X	7	X	32,183.69	=	$ 39,581.88
4th draw	1.03	Y^X	9	X	32,183.69	=	$ 41,992.42

Check the amounts by doing the following

Fund started with	$100.000.00
After 2 yrs before 1st child is paid fund is	
then worth 100,000 X $(1.08)^2$	116,640.00
The first child is paid	34,143.68
Balance	82,496.32
After 2 more yrs. the fund is worth 82,496.32 X $(1.08)^2$	96,223.71
The 2nd child is paid	36,223.03
Balance	60,000.68
After 3 more yrs. the fund is worth 60,000.68 X $(1.08)^3$	75,583.58
The 3rd child is paid	39,581.88
Balance	36,001.70
After 2 more yrs. the fund is worth 36,001.70 X $(1.08)^2$	41,992.38
The 4th child is paid	41,992.42
	- .04

Money - Funding

Present Value of a Future Amount

The term used to express what a dollar is valued at today verses a future day.

Example- You are going to inherit $ 500,000 in 10 years. What is the inheritance worth today (present buying power). Estimated inflation rate will probably average 3.5 percent a year.

$$PV = \frac{\text{Future Amount}}{(1 + \text{Rate})^{\text{Number of years}}}$$

$$PV = \frac{\$ 500,000}{(1 + .035)^{10}}$$

$$PV = \frac{\$ 500,000}{(1 .035)^{10}}$$

Calculate 1.035 Y^X 10 = 1.4105988

$$PV = \frac{\$ 500,000}{1.4105988} = 354,459.41$$

PV = $ $ 354,459.41 would buy you the same amount of goods today as $500,000 would after 10 years of inflation at 3.5% per year.

Another application of-
Present Value of Future Amount

How much money do you have to invest today to get a determined amount at a future day at a specific interest rate?

Example -The Future Amount needed, 10 years from now is $ 20,000. If interest is compounded yearly at 7 %, how much would be needed to be deposited today?

(continued)

Present Value (continued)

$$PV = \frac{\$\,20{,}000}{(\,1 + .07\,)^{10}}$$

$$PV = \frac{20{,}000}{1.07^{\,10}}$$

Calculate 1.07 $\boxed{Y^X}$ *10* $\boxed{=}$ *1.961514*

$$PV = \frac{20{,}000}{1.9671514}$$

$$PV = \$\,10{,}166.99$$

> The Present Value of $ 20,000 ten years from now is $10,166.99

Example -%. The future amount needed 10 years from now is $ 20,000 . The interest is compounded <u>monthly</u> at an annual interest rate of 7%. How much has to be deposited today?

$$PV = \frac{\text{Future Amount Needed}}{1 + \dfrac{\text{Rate}}{\text{Compound periods per year}}^{\substack{\text{Number of Compound Periods}\\ \text{During Program}}}}$$

$$PV = \frac{20{,}000}{\left(1 + \dfrac{.07}{12}\right)^{10 \text{ years X 12 months}}}$$

0.07 divided by 12 = 0.0058333

$$PV = \frac{20{,}000}{1 + .0058333^{\,120}}$$

$$PV = \frac{20{,}000}{1.0058333^{\,120}}$$

calculate 1.0058333 $\boxed{Y^x}$ *120* $\boxed{=}$ *2.0096614*

$$PV = \frac{20{,}000}{2.0096614} \qquad PV = 9{,}951.93$$

Present Value of One (continued)

A Different Way of Figuring It
In addition to the previous formulas there is another way of determining the Present Value of a dollar in the future.

So that you will understand this mathematically, we will do a problem without a calculator with a $\boxed{Y^X}$ key and a $\boxed{+ -}$ key.

Example - What is the present value of one, at 7 % interest compounded annually for 5 years?

Step 1 Divide 1 by 1.07 = .9345794
Step 2 Divide .9345794 by 1.07 = .8734387
Step 3 Divide .8734387 by 1.07 = .8162979
Step 4 Divide .8162979 by 1.07 = .7628952
Step 5 Divide .7628952 by 1.07 = .7129862

.7129862 is the present value of $ 1.00 5 years from now at 7 % interest compounded annually.

We can now multiply it times any dollar amount to figure its present value.

Example- What is the present value of $ 2,000 5 years from now at 7 % interest compounded annually? Stated another way- How much do we have to invest today to have $ 2,000, 5 years from now at 7 % interest compounded annually?

$ 2,000 X .7129862 = $ 1,425.97

With a calculator that has $\boxed{Y^X}$ and $\boxed{\pm}$ keys.

1.07 $\boxed{Y^X}$ 5 $\boxed{\pm}$ = .7129862

$ 2,000 X .7129862 = $ 1,425.97

Present Value (continued)

Example - What is the present value of $ 13,500
10 years from now, compounded monthly
at 8 % interest.

$$PV = 13,500 \times \left(1 + \frac{.08}{12} \right)^{-10 \text{ years} \times 12}$$

$$PV = 13,500 \times (1.0066667)^{-120}$$

using a calculator do the following

13,500 X 1.0066667 $\boxed{Y^X}$ 120 $\boxed{\pm}$ = $ 6,082.04

PV = $ 6,082.04

Money - Loans

Rule of 78

This is a way of structuring out a loan, to determine
how much interest and principal has been paid during
the loan period of one year . It can be applied to longer
loan periods as you will see.

Example - You borrow $ 3,500 at 8% interest for
12 months. You will have use of the complete
3,500 for only one month, then you have to give
some of it back in the form of a payment and
every month thereafter. The interest on the
loan is $ 280 (3,500 X .08 = 280). The total
amount of the loan is then $ 3,780. Divide the
loan into 12 equal payments of $ 315.

When you add up the numbers

12+11+10+9+8+7+6+5+4+3+2+1 they equal 78

	Make fractions using each month and 78		Now change the fractions to decimals	Multiply times the interest of the loan		Portion of each payment used to pay off the interest
1	12/78	=	.1538462	X	280 =	43.08
2	11/78	=	.1410256	X	280 =	39.49
3	10/78	=	.1282051	X	280 =	35.90
4	9/78	=	.1153846	X	280 =	32.31
5	8/78	=	.1025641	X	280 =	28.72
6	7/78	=	.0897436	X	280 =	25.13
7	6/78	=	.0769231	X	280 =	21.54
8	5/78	=	.0641026	X	280 =	17.95
9	4/78	=	.0512821	X	280 =	14.36
10	3/78	=	.0384615	X	280 =	10.77
11	2/78	=	.025641	X	280 =	7.18
12	1/78	=	.0128205	X	280 =	3.59

Please note the totals   100.0 $ 280.02

Now that we have determined how much of each payment goes
to pay off the interest, we can now determine how much goes
to pay off the principal.

Rule of 78 (continued)

	Payment		Interest payment per month		Portion of each Payment applied to principal
1	315	minus	43.08	=	271.92
2	315	-	39.49	=	275.51
3	315	-	35.90	=	279.10
4	315	-	32.31	=	282.69
5	315	-	28.72	=	286.28
6	315	-	25.13	=	289.87
7	315	-	21.54	=	293.46
8	315	-	17.95	=	297.05
9	315	-	14.36	=	300.64
10	315	-	10.77	=	304.23
11	315	-	7.18	=	307.82
12	315	-	3.59	=	311.41

$ 3,499.98 or $ 3,500.00

Please note the total applied to the principal equals the amount borrowed.

Now that we know how much is applied to the principal, we can figure the amount of money necessary to pay off the loan at any given time.

	Loan balance before each payment		Portion of each applied to the principal		Loan balance after each payment
1	3780.00	-	271.92	=	3508.08
2	3508.08	-	275.51	=	3232.57
3	3232.57	-	279.10	=	2953.47
4	2953.47	-	282.69	=	2670.78
5	2670.78	-	286.28	=	2384.50
6	2384.50	-	289.87	=	2094.63
7	2094.63	-	293.46	=	1801.17
8	1801.17	-	297.05	=	1504.12
9	1504.12	-	300.64	=	1203.48
10	1203.48	-	304.23	=	899.25
11	899.25	-	307.82	=	591.43
12	591.43	-	311.41	=	280.02

Rule of 78 (continued)

As you can see on the previous page, the more payments you make, the amount that you owe decreases at an accelerated pace. Note that after 6 months, half of the payments made, you owe $ 2,094.63. You have paid out $ 315 X 6 = $ 1,890.00.

Pay Off of a Rule of 78 Loan

Instead of making a table as we have done, to determine the pay off of a Rule of 78 loan, the following formula can be used.

$$\text{Pay off at any given time} = \text{Amount of Loan} - \left[\text{Monthly payment amount} \times \text{Number of payments made} - \left(\frac{SN}{SD} \times \text{Amount of Interest} \right) \right]$$

Example- We will use the previous example. What is the pay off after 6 payments of a $3,780 loan with 12 payments of 315.00 ? The loan includes $280.00 interest.

$$\text{Pay off after 6 payments} = 3,780 - \left[315 \times 6 - \left(\frac{SN}{SD} \times 280 \right) \right]$$

SN = 12+11+10+9+8+7 = 57   This is the sum of the payment number of those payments not made.

or use the following formula

$$SN = \frac{\text{Payment}}{\text{Number}} \times \frac{\text{Total number of Payments of Loan}}{} - \left[\frac{\text{Payment Number} \left(\text{Payment Number} + 1 \right)}{2} \right] + \frac{\text{Payment Number}}{}$$

$$SN = 6 \times 12 - \left[\frac{6 \ (\ 6 + 1 \ )}{2} \right] + 6$$

$$SN = 72 - \frac{42}{2} + 6 \quad SN = 72 - 21 + 6 = 57$$

Rule of 78 (continued)

Total the payment numbers for each monthly payment.

SD = 12+11+10+9+8+7+6+5+4+3+2+1 = 78

or use the following formula

$$SD = \frac{\frac{\text{Total number of Payments of Loan}}{} \times \left(\frac{\text{Total number of Payments of Loan}}{} + 1\right)}{2}$$

$$SD = \frac{12 \times (12 + 1)}{2} \qquad SD = 78$$

We will now complete the formula

$$\text{Pay off after 6 payments} = 3{,}780 - \left[315 \quad \times \quad 6 \quad - \left(\frac{57}{78} \times 280\right)\right]$$

$$\text{Pay off after 6 payments} = 3{,}780 - \left[315 \times 6 - (0.7307692 \times 280)\right]$$

$$\text{Pay off after 6 payments} = 3{,}780 - \left[315 \times 6 - (204.61538)\right]$$

$$\text{Pay off after 6 payments} = 3{,}780 - (1890 - 204.61538)$$

$$\text{Pay off after 6 payments} = 3{,}780 - 1685.38$$

$$\text{Pay off after 6 payments} = \$2{,}094.62$$

Money - Loans

Payment Amount for Short Term Loans

Example - What would the monthly payment be for a loan of $ 3,000 at 7 % interest to be paid back in 8 months?

$$\frac{\text{Loan}}{\text{Payment}} = \frac{\text{Amount}}{\text{Loaned}} \quad X \quad \frac{\left[\frac{\text{Rate}}{12} \; X \left(1 + \frac{\text{Rate}}{12}\right)^{\text{Months}}\right]}{\left[\left(1 + \frac{\text{Rate}}{12}\right)^{\text{Months}}\right] - 1}$$

$$\frac{\text{Loan}}{\text{Payment}} = 3,000 \quad X \quad \frac{\left[\frac{.07}{12} \; X \left(1 + \frac{.07}{12}\right)^{8}\right]}{\left[\left(1 + \frac{.07}{12}\right)^{8}\right] - 1}$$

.07 divided by 12 = .0058333

$$\frac{\text{Loan}}{\text{Payment}} = 3,000 \quad X \quad \frac{\left[.0058333 \; X \; (1.0058333)^{8}\right]}{\left[(1.0058333)^{8}\right] - 1}$$

Calculate $1.0058333 \boxed{Y^X} \; 8 \boxed{=} 1.0476306$

$$\frac{\text{Loan}}{\text{Payment}} = 3,000 \quad X \frac{.0058333 \; X \; 1.0476306}{1.0476306 \; - 1}$$

.0058333 X 1.0476306 = .0061111

$$\frac{\text{Loan}}{\text{Payment}} = 3,000 \quad X \quad \frac{.0061111}{.0476306}$$

$$\frac{\text{Loan}}{\text{Payment}} = 3,000 \; X \; .1283029 \; = \; \$ \; 384.91$$

Payment Amount for Loans or Mortgages

This is the same formula as for Mortgage Payments.

Example - A person is buying a car for $ 18,350 and is getting a loan at 9 % interest for 5 years. What is the monthly payment?

$$\text{Payment} = \frac{\text{Loan}}{\text{Amount}} \times \left\{ \frac{\left[\left(\frac{\text{Rate}}{12}\right) \times \left(1 + \frac{\text{Rate}}{12}\right)^{\text{Number of payments}}\right]}{\left[\left(1 + \frac{\text{Rate}}{12}\right)\right]^{\text{Number of payments}} - 1} \right\}$$

$$\text{Payment} = \$\,18{,}350 \times \left\{ \frac{\left[\left(\frac{.09}{12}\right) \times \left(1 + \frac{.09}{12}\right)^{5 \text{ years X } 12 \text{ months}}\right]}{\left[\left(1 + \frac{.09}{12}\right)^{60}\right] - 1} \right\}$$

.09 divided by 12 = .0075
then add 1 = 1.0075

$$\text{Payment} = \$\,18{,}350 \times \left\{ \frac{\left[(.0075) \times (1.0075)^{60}\right]}{\left[(1.0075)^{60}\right] - 1} \right\}$$

calculate 1 .0075 $\boxed{Y^x}$ 60 $\boxed{=}$ 1.565681

$$\text{Payment} = \$\,18{,}350 \times \frac{[.0075 \text{ X } 1.565681]}{[1.565681 - 1]}$$

.0075 X 1.565681 = .0117426

$$\text{Payment} = \$\,18{,}350 \times \frac{.0117426}{.565681}$$

Payment = $ 18,350   X .0207584 = 380.92

Payment = $ 380.92

Money - Loans

Mortgage Payment - Monthly

$$MP = \frac{Amount}{Mortgaged} \quad X \quad \frac{\dfrac{Rate}{12} \; X \left(1 + \dfrac{Rate}{12}\right)^{Months \; X \; Years}}{\left(1 + \dfrac{Rate}{12}\right)^{Months \; X \; Years} - 1}$$

Example - What is the monthly mortgage payment for $ 220,000 at 7% interest for 30 years?

$$MP = 220,000 \quad X \quad \frac{\dfrac{.07}{12} \; X \left[\left(1 + \dfrac{.07}{12}\right)^{360}\right]}{\left[\left(1 + \dfrac{.07}{12}\right)^{360}\right] - 1}$$

.07 divided by 12 = .0058333
then add the 1 equaling 1.0058333

$$MP = 220,000 \quad X \quad \frac{.0058333 \; X \; (1.0058333)^{360}}{\left[(1.0058333)^{360}\right] - 1}$$

Calculate 1.0058333 $\boxed{Y^x}$ *360* $\boxed{=}$ *8.1164974*

$$MP = 220,000 \quad X \quad \frac{.0058333 \; X \; 8.1164974}{8.1164974 - 1} \qquad \textit{subtract}$$

$$MP = 220,000 \quad X \quad \frac{.0058333 \; X \; 8.1164974}{7.1164974} \qquad \textit{multiply}$$

$$MP = 220,000 \quad X \quad \frac{.047346}{7.1164974} \qquad \textit{divide}$$

$$MP = 220,000 \quad X \quad .006653 = \$ 1463.65$$

Loan or Mortgage Payment - Monthly

You have seen in the previous formulas, the right side of the formula is based on time and interest rate. Once you have that number, (the multiplier) you can multiply it times any loan amount and determine the monthly payment. The following table supplies the multiplier.

Example- What is the monthly loan payment, for $250,000 at 12.25% interest for 5 years.

Turn to Page 9:60 and go down the left hand Interest Column, to 12.25. Then go across to the 5 year column to the number .02237099. Multiply this number times the loan amount, to determine monthly payment.

.02237099 X 250,000 = $5, 592.75 is the monthly payment

Short Term Loans

Interest Rate	Years 2	3	4
3.50	.04320272	.02930208	.02235600
3.75	.0433174	.02941290	.02246736
4.00	.04342492	.02952399	.02257905
4.25	.04353628	.02963533	.02269110
4.5	.04364781	.02974692	.02280349
4.75	.04375951	.02985878	.02291622
5.00	.04387139	.02997090	.02302929
5.25	.04398344	.03008327	.02314271
5.5	.04409566	.03019590	.02325648
5.75	.04420805	.03300879	.02337058
6.00	.04432061	.03042194	.02348503
6.25	.04443334	.03053534	.02359982
6.50	.04454625	.03064900	.02371495
6.75	.04465933	.03076292	.02383043
7.00	.04477258	.03087710	.02394624
7.25	.04488600	.03099153	.02406240
7.50	.04499959	.03110622	.02417890
7.75	.04511336	.03122116	.02429574
8.00	.04522729	.03133637	.02441292
8.25	.04534140	.03145182	.02453044
8.50	.04545567	.03156754	.02464830
8.75	.04557012	.03168351	.02476650
9.00	.04568474	.03179973	.02488504
9.25	.04579953	.03191621	.02500392

Money - Loans

Short Term Loans
(continued)

Interest Rate	Years		
	2	**3**	**4**
9.50	.04591449	.03203295	.02512314
9.75	.04602962	.03214994	.02524269
10.00	.04614493	.03226719	.02536258
10.25	.04626040	.03235469	.02548281
10.50	.04637604	.03250244	.02560388
10.75	.04649185	.03262045	.02572428
11.00	.04660784	.03273872	.02584552
11.25	.04672399	.03285723`	.02596710
11.50	.04684032	.03297601	.02608901
11.75	.04695681	.03309503	.02621125
12.00	.04707347	.03321431	.02633384
12.25	.04719031	.03333384	.02645675
12.50	.04730731	.03345363	.02658000
12.75	.04742448	.03357366	.02670358
13.00	.04754782	.03369395	.02682750
13.25	.04765933	.03381449	.02695174
13.50	.04777701	.03393529	.02707632
13.75	.04789486	.03405633	.02720123
14.00	.04801288	.03417763	.02732648
14.25	.04813107	.03429918	.02745205
14.50	.04824943	.03442098	.02757795
14.75	.04836795	.03454303	.02770419
15.00	.04848665	.03466533	.02783075
15.25	.04860551	.03478788	.02795764
15.50	.04872455	.03491068	.02808486
15.75	.04884374	.03503373	.02821241
16.00	.04896311	.03515703	.02834028
16.25	.04908265	.03528058	.02846848
16.50	.04920235	.03540438	.02859701
16.75	.04932222	.03552843	.02872586
17.00	.04944226	.03565273	.02885504
17.25	.04956247	.03577727	.02898455
17.50	.04968285	.03590207	.02911437
17.75	.04980339	.03602711	.02924453
18.00	.04992410	.03615240	.02937500
18.25	.05004498	.03627793	.02950580
18.50	.05016603	.03640371	.02963692
18.75	.05028724	.03652974	.02976836
19.00	.05040862	.03665602	.02990012
19.25	.05053016	.03678254	.03003220
19.50	.05065188	.03690931	.03016460
19.75	.05077376	.03703632	.03029732
20.00	.05089580	.03716358	.03043036
20.25	.05101802	.03729109	.03056372
20.50	.05114039	.03741884	.03069739
20.75	.05126294	.03754683	.03083139
21.00	.05138565	.03767507	.03096569
21.25	.05150853	.03780355	.03110032

Short Term Loans
(continued)

	Years		
	2	**3**	**4**
21.5	.05163157	.03793227	.03123526
21.75	.05175478	.03806124	.03137051
22.00	.05187815	.03819045	.03150608
22.25	.05200169	.03831991	.03164196
22.50	.05212540	.03844960	.03177815
22.75	.05224927	.03857954	.03191466
23.00	.05237331	.03870972	.03205147

	5	**6**	**10**
3.50	.01819174	.01541840	.0098859
3.75	.01830392	.01553153	.01000612
4.00	.01841652	.01564518	.01012451
4.25	.01852956	.01575935	.01024375
4.50	.01864302	.01587403	.01036384
4.75	.01875691	.01598922	.01048477
5.00	.01887123	.01610493`	.01060655
5.25	.01898598	.01622115	.01072917
5.50	.01910116	.01633789	.01085263
5.75	.01921677	.01645513	.01097692
6.00	.01933280	.01657289	.01110205
6.25	.01944925	.01669115	.01122801
6.50	.01956615	.01680993	.01135480
6.75	.01968346	.01692921	.01148241
7.00	.01980120	.01704901	.01161085
7.25	.01991936	.01716931	.01174010
7.50	.02003795	.01729011	.01187018
7.75	.02015696	.01741142	.01200106
8.00	.02027639	.01753324	.01213276
8.25	.02039625	.01765556	.01226526
8.50	.02051653	.01777838	.01239857
8.75	.02063723	.01790171	.01253268
9.00	.02075836	.01802554	.01266758
9.25	.02087990	.01814986	.01280327
9.50	.02100186	.01827469	.01293976
9.75	.02112424	.01840002	.01307702
10.00	.02124704	.01852584	.01321507
10.25	.02137026	.01865216	.01335390
10.50	.02149390	.01877897	.01349350
10.75	.02161795	.01890628	.01363387
11.00	.02174242	.01903408	.01388500
11.25	.02186731	.01916237	.01391689
11.50	.02199261	.01929116	.01405954
11.75	.02211835	.01942043	.01420295
12.00	.02224445	.01955019	.01434709

Money- Loans

Short Term Loans
(continued)

Interest Rate	Years		
	5	6	10
12.25	.02237099	.01968044	.01449199
12.50	.02249794	.01981118	.01463762
12.75	.02262530	.01994240	.01478398
13.00	.02275307	.02007411	.01493107
13.25	.02288126	.02020629	.01507889
13.50	.02300985	.02033896	.01522743
13.75	.02313884	.02047211	.01537668
14.00	.02326825	.02060574	.01552664
14.25	.02339806	.02073985	.01567731
14.50	.02352828	.02087443	.01582868
14.75	.02365890	.02100948	.01598074
15.00	.02378993	.02114501	.01613350
15.25	.02392136	.02128102	.01628693
15.50	.02405319	.02141749	.01644105
15.75	.02418542	.02155443	.01659585
16.00	.02431806	.02169184	.01675131
16.25	.02445109	.02182972	.01690744
16.50	.02458452	.02196806	.01706423
16.75	.02471835	.02210686	.01722167
17.00	.02485258	.02224613	.01737977
17.25	.02498720	.02238586	.01753850
17.50	.02512221	.02252605	.01769788
17.75	.02525762	.02266669	.01785788
18.00	.02539343	.02280779	.01801852
18.25	.02552962	.02294935	.01817978
18.50	.02566621	.02309135	.01834165
18.75	.02580319	.02323381	.01850414
19.00	.02594055	.02337672	.01866724
19.25	.02607830	.02352008	.01883093
19.50	.02621645	.02366388	.01899522
19.75	.02635497	.02380813	.01916010
20.00	.02649388	.02395283	.01932557
20.25	.02663318	.02409796	.01949161
20.50	.02677286	.02424353	.01965823
20.75	.02691292	.02438955	.01982542
21.00	.02705336	.02453600	.01999317
21.25	.02719418	.02468288	.02016147
21.50	.02733538	.02483020	.02033033
21.75	.02747696	.02497795	.02049974
22.00	.02761891	.02512613	.02066969
22.25	.02776124	.02527473	.02084017
22.50	.02790395	.02542377	.02101118
22.75	.02804702	.02557323	.02118272
23.00	.02819047	.02572311	.02135478

Long Term Loans

Interest Rate	Years		
	15	20	30
5.00	.00790794	.00659956	.00536822
5.25	.00803878	.00673844	.00552204
5.50	.00817083	.00687887	.00567789
5.75	.00830410	.00702084	.00583573
6.00	.00843857	.00716431	.00599551
6.25	.00857423	.00730928	.00615717
6.50	.00871107	.00745573	.00632068
6.75	.00884909	.00760364	.00648598
7.00	.00898828	.00775299	.00665302
7.25	.00912863	.00790376	.00682176
7.50	.00927012	.00805593	.00699215
7.75	.00941276	.00820949	.00716412
8.00	.00955652	.00836440	.00733765
8.25	.00970140	.00852066	.00751267
8.50	.00984740	.00867823	.00768913
8.75	.00999449	.00883711	.00786700
9.00	.01014267	.00899726	.00804623
9.25	.01029192	.00915867	.00822675
9.50	.01044225	.00932131	.00840854
9.75	.01059363	.00948517	.00859154
10.00	.01074605	.00965022	.00877572
10.25	.01089951	.00981643	.00896101
10.50	.01105399	.00998380	.00914739
10.75	.01120948	.01015229	.00933481
11.00	.01136597	.01032188	.00952323
11.25	.01152345	.01049256	.00971261
11.50	.01168190	.01066430	.00990291
11.75	.01184131	.01083707	.01009410
12.00	.01200168	.01101086	.01028613
12.25	.01216299	.01118565	.01449199
12.75	.01248837	.01153812	.01086693
13.00	.01265242	.01171576	.01106200
13.25	.01281736	.01189431	.01125774
13.5	.01298319	.01207375	.01145412
13.75	.01314987	.01225405	.01165113
14.00	.01331741	.01243521	.01184872

Money - Loans

Pay Off - Midterm of a Loan or Mortgage (formula on next page)

In the example $ 18,350 was borrowed at 9 % interest with monthly payments of $380.92 for 5 years. If the person borrowing the money wanted to pay off the loan after 18 months what would that amount be?

(The first part of the equation is to determine how much the money loaned would have made, had it been put in a bank earning interest for the time used.)

The left side of the equation will be worked first.

The rate of interest 9% is changed to .09 and is then divided by 12 the number of compounding periods in one year, equaling .0075. 1 is added to it equaling 1.0075 and then is multiplied times itself 18 times.

Calculate   1.0075 $\boxed{Y^x}$ 18 $\boxed{=}$ 1.1439604

The Loan amount $18,350 is then multiplied by 1.1439604 equaling $ 20,991.67.

Now do the right side of the minus sign. *(The same as figuring how much would be deposited if the payments were put in the bank)*

Again the rate of interest 9% is changed to .09 and is then divided by 12 the number of compounding periods in one year, equaling .0075. 1 is added to it equaling 1.0075 and then is multiplied times itself 18 times.

Calculate   1.0075 $\boxed{Y^x}$ 18 $\boxed{=}$ 1.1439604

The rate .09 is also divided by 12 equaling .0075 and will be used to divide into the upper part of the equation.
1 is subtracted from 1.1439604 equaling .1439604 and is divided by .0075 equaling 19.194719.
19.194719 is then multiplied by the amount of each payment $380.92 equaling $7,311.65.

Pay Off =   20,991.67 - 7,311.65 = $ 13,680.02
Please note that the Equity on the loan = $18,350 - 13680.02 = $ 4,669.98

Pay Off - Midterm of a Loan or Mortgage

The value of the loan, had it been put in the bank earning interest	−	The amount of money you would have, had you put the payments in the bank earning interest

Example : The amount borrowed was $ 18,350 at 9% interest for 5 years. After 18 months of making $380.92 payments per month , how much would it take to payoff the loan?

$$\text{Loan Amount} \times \left(1 + \frac{\text{rate}}{12}\right)^{\text{No. of payments made}} - \text{Payment} \times \frac{\left(1 + \frac{\text{rate}}{12}\right)^{\text{No. of payments made}} - 1}{\frac{\text{rate}}{12}}$$

$$\text{Pay off} = 18,350 \times \left[\left(1 + \frac{.09}{12}\right)^{18}\right] - \left\{380.92 \times \frac{\left[\left(1 + \frac{.09}{12}\right)^{18}\right] - 1}{\frac{.09}{12}}\right\}$$

$$\text{Pay off} = \left[18,350 \times 1.0075^{18}\right] - \left\{380.92 \times \frac{\left[1.0075^{18}\right] - 1}{.0075}\right\}$$

calculate 1.0075 Y^X key 18 = 1.1439604

$$\text{Pay off} = \left[18,350 \times 1.1439604\right] - \left\{380.92 \times \frac{.1439604}{.0075}\right\}$$

$$\text{Pay off} = \left[18,350 \times 1.1439604\right] - \left[380.92 \times 19.194719\right]$$

$$\text{Pay off} = 20,991.67 - 7,311.65 = 13,680.02$$

$$\text{Pay off} = \$ 13,680.02$$

The equity or the amount paid on the loan is

$$18,350 - 13,680.02 = \$4,669.98$$

Determining Pay-Off on an Unstructured Loan (the payments are a predetermined amount)

(the formula is on the next page)

Example- A Customer has agreed to pay 20 payments of $1,000 on a $20,000 loan. The interest is 15 % and will be paid with the last payment. After making 14 payments the customer wants to pay off the loan.

The interest rate of .15 is divided by 12 equaling 0.0125. 1 is added to it equaling 1.0125. A calculator is used to multiply 1.0125 times itself 14 times (the number of payments made) equaling 1.1899548

Calculate 1.0125 Y^X key 14 = 1.1899548

In the right side of the equation 1 is subtracted from 1.1899548 equaling 0.1899548

In the left side of the equation 1.1899548 is multiplied by 20,000 equaling 23,799.10. In the right side 0.1899548 is divided by 0.0125 equaling 15.196384

The monthly payment of $1,000 is then multiplied by 15.196384 equaling 15,196.38.

15,196.38 is then subtracted from 23,799.10 equaling $ 8,602.72 the amount needed to pay off the loan.

Please note that the $ 23,799.10 would be the amount the person who loaned the money would have, had they put their $ 20,000 in the bank earning 15% interest, compounded monthly. The $ 15,196.38 is the amount the customer would have, had they put their payments in a savings account earning 15% interest compounded monthly.

To determine the interest payment of this type of loan
please see page 9 : 11

Determining Pay-Off on Unstructured Loan

(this is the same as a regular loan, except the payments are rounded off to a set amount.)

Example- $ 20,000 is loaned at 15 % interest and 20 monthly payments of $ 1,000 are to be made. The interest will be paid after the last payment. What is the pay-off after 14 payments have been made?

$$\text{Pay-Off} = \frac{\text{Amount}}{\text{Loaned}} \left(1 + \frac{\text{rate}}{12}\right)^{\text{number of payments made}} - \frac{\text{Monthly}}{\text{Payment}} \cdot \frac{\left(1 + \frac{\text{rate}}{12}\right)^{\text{number of payments made}} - 1}{\frac{\text{rate}}{12}}$$

$$\text{Pay-Off} = \left\{ 20,000 \times \left(1 + \frac{.15}{12}\right)^{14} \right\} - \left\{ 1,000 \times \frac{\left[\left(1 + \frac{.15}{12}\right)^{14}\right] - 1}{\frac{.15}{12}} \right\}$$

.15 divided by 12 = 0.0125 then add the 1

$$\text{Pay-Off} = \left\{ 20,000 \times \left(1.0125\right)^{14} \right\} - \left\{ 1,000 \times \frac{\left(1.0125\right)^{14} - 1}{0.0125} \right\}$$

calculate 1.0125 Y^x key 14 = 1.1899548

$$\text{Pay-Off} = \left(20,000 \times 1.1899547\right) - \left\{ 1,000 \times \frac{1.1899547 - 1}{0.0125} \right\}$$

$$\text{Pay-Off} = \left(20,000 \times 1.1899547\right) - \left\{ 1,000 \times \frac{0.1899547}{0.0125} \right\}$$

$$\text{Pay-Off} = \left(23,799.09\right) - \left\{ 1,000 \times 15.196376 \right\}$$

$$\text{Pay-Off} = 23,799.09 - 15,196.38$$

$$\text{Pay-Off} = \$ 8,602.71$$

Mortgage or Loan Amount Unknown - Monthly payment is known

(formula on next page)

Step 1

The rate of interest 7% is changed to .07 and is divided by 12 the number of compounding periods in a year, equally .0058333. This is done for the upper (numerator) part of the equation and the lower (denominator) part of the equation. For the upper part 1 is added equaling 1.005833. The total number of payments is figured 12 months times 30 years equaling 360.

1.0058333 is then multiplied times itself 360 times.

Calculate 1.0058333 $\boxed{Y^x}$ 360 $\boxed{=}$ 8.1164006
1 is subtracted equaling 7.116406
7.116406 is then divided by .0058333 equaling 1,219.9623

Step 2

The answer from Step 1 1,219.9623 is then divided by Step 2

The lower part of the equation of Step 2 is the same as a part of the upper portion of Step 1. As follows,

The rate of interest 7% is changed to .07 and is divided by 12 the number of compounding periods in a year, equally .0058333. 1 is then added equaling 1.005833. The total number of payments is figured 12 months times 30 years equaling 360.
1.0058333 is then multiplied times itself 360 times.

Calculate 1.0058333 $\boxed{Y^x}$ 360 $\boxed{=}$ 8.1164006

1,219.977954 is then divided by 8.1164006 equaling 150.3102

$ 1,465 times 150.3102 equals $ 220,204.44

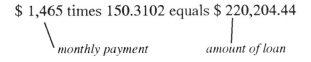

monthly payment   *amount of loan*

Mortgage or Loan Amount Unknown - Monthly payment is **Known**

(the instructions are on the previous page)

Step 1

$$\frac{\text{Loan}}{\text{Amount}} = \frac{\text{Monthly}}{\text{Payment}} \times \frac{\left(1 + \dfrac{\text{Rate}}{12}\right)^{\text{Number of Payments}} - 1}{\dfrac{\text{Rate}}{12}} = \frac{\text{Step 1}}{\text{Divided by Step 2}}$$

Step 2

$$\frac{\text{Loan}}{\text{Amount}} = \frac{\text{Monthly}}{\text{Payment}} \times \frac{\text{Step 1}}{\left(1 + \dfrac{\text{Rate}}{12}\right)^{\text{Number of payments}}}$$

Example - If you have $1,465.00 available for house payments how big of a Mortgage could you have with a 7% 30 year loan?

Step 1

$$\frac{\left[\left(1 + \dfrac{.07}{12}\right)^{12 \times 30}\right] - 1}{\dfrac{.07}{12}} = \frac{\left[(1.0058333)^{360}\right] - 1}{.0058333}$$

$$\frac{8.1164974 \ -1}{.0058333} = \frac{7.1164974}{.0058333} = 1219.977954$$

Step 2

$$\frac{\text{Loan}}{\text{Amount}} = \$1465 \times \frac{1219.977954}{\text{Step 2}} = \frac{1219.977954}{\left(1 + \dfrac{\text{Rate}}{12}\right)^{\text{Number of payments}}}$$

$$\frac{\text{Loan}}{\text{Amount}} = \$1465 \times \frac{1219.977954}{\left(1 + \dfrac{.07}{12}\right)^{12 \times 30}} = \frac{1219.977954}{(1.0058333)^{360}}$$

$$\frac{\text{Loan}}{\text{Amount}} = \$1,465 \times \frac{1219.977954}{8.1164006} = 150.3102$$

$$\frac{\text{Loan}}{\text{Amount}} = \$1,465 \times 150.3102 = 220,204.44 \quad \text{the amount that can be mortgaged}$$

Money - Loans

Car or Loan Amount Unknown - Monthly payment is Known

(this is the same formula for mortgages)

Step 1

$$\frac{\text{Loan}}{\text{Amount}} = \frac{\text{Monthly}}{\text{Payment}} \times \frac{\left(1 + \dfrac{\text{Rate}}{12}\right)^{\text{Number of Payments}} - 1}{\dfrac{\text{Rate}}{12}} = \frac{\text{Step 1}}{\text{Divided by Step 2}}$$

Step 2

$$\frac{\text{Loan}}{\text{Amount}} = \frac{\text{Monthly}}{\text{Payment}} \times \frac{\text{Step 1}}{\left(1 + \dfrac{\text{Rate}}{12}\right)^{\text{Number of payments}}}$$

Example - If you have $350 available for car payments how expensive of a car could you have with a 13%, 5 year loan?

Step 1

$$\frac{\left[\left(1 + \dfrac{.13}{12}\right)^{12 \times 5}\right] - 1}{\dfrac{.13}{12}} = \frac{\left[(1.0108333)^{60}\right] - 1}{.0108333}$$

$$\frac{1.9088565 - 1}{.0108333} = \frac{0.9088565}{.0108333} = 83.894707$$

Step 2

$$\frac{\text{Loan}}{\text{Amount}} = \$350 \times \frac{83.894707}{\text{Step 2}} = \frac{83.894707}{\left(1 + \dfrac{\text{Rate}}{12}\right)^{\text{Number of payments}}}$$

$$\frac{\text{Loan}}{\text{Amount}} = \$350 \times \frac{83.894707}{\left(1 + \dfrac{.13}{12}\right)^{12 \times 5}} = \frac{83.894707}{(1.0108333)^{60}}$$

$$\frac{\text{Loan}}{\text{Amount}} = \$350 \times \frac{83.894707}{1.9088565} = 43.950243$$

$$\frac{\text{Loan}}{\text{Amount}} = \$350 \times 43.950243 = \$ 15,382.59$$

the total amount that can be financed

Table A

Amount of 1 at Compound Interest

COMPOUNDED
YEARLY, TWICE YEARLY, QUARTERLY

$$\left(1 + \frac{\text{Interest Rate}}{\text{Number of compound periods in 1 year}} \right)^{N}$$

N = number of compound periods during program

Example - 8 % interest compounded yearly for 10 years

$$\left(1 + \frac{.08}{1} \right)^{10} = \left(1 + .08 \right)^{10} = \left(1.08 \right)^{10} = 2.158925$$

N	1.05	1.06	1.07	1.08
2	1.1025	1.1236	1.1449	1.1664
3	1.157625	1.191016	1.225043	1.259712
4	1.215506	1.262477	1.310796	1.360489
5	1.276282	1.338226	1.402552	1.469328
6	1.340096	1.418519	1.500730	1.586874
7	1.407100	1.503630	1.605782	1.713824
8	1.477455	1.593848	1.718186	1.850930
9	1.551328	1.689479	1.838459	1.999005
10	1.628895	1.790848	1.961514	2.158925
12	1.795856	2.012197	2.252192	2.518170
14	1.979932	2.260904	2.578534	2.937194
16	2.182875	2.540352	2.952164	3.425943
18	2.406619	2.854339	3.379932	3.996020
20	2.653298	3.207136	3.869685	4.660957
22	2.925261	3.603537	4.430402	5.436540
24	3.225100	4.048935	5.072367	6.341181
25	3.386355	4.291871	5.427433	6.848475
30	4.321942	5.743491	7.612255	10.06266
35	5.516015	7.686087	10.67658	14.78534
40	7.039989	10.28572	14.97446	21.72452
45	8.985008	13.76461	21.00245	31.92045
50	11.46740	18.42015	29.45703	46.90161
55	14.63563	24.65032	41.31500	68.91386
60	18.67919	32.98769	57.94643	101.2571

Table A
Amount of 1 at Compound Interest

COMPOUNDED
YEARLY, TWICE YEARLY, QUARTERLY
(continued)

N	1.09	1.10	1.11	1.12
2	1.188100	1.21000	1.232100	1.254400
3	1.295029	1.33100	1.367631	1.404928
4	1.411582	1.46410	1.518070	1.404928
5	1.538624	1.61051	1.685058	1.762342
6	1.677100	1.77156	1.870415	1.973823
7	1.828039	1.94872	2.076160	2.210681
8	1.992563	2.14359	2.304538	2.475963
9	2.171893	2.35795	2.558037	2.773079
10	2.367364	2.593742	2.839421	3.105848
12	2.812665	3.138428	3.498451	3.895976
14	3.341727	3.797498	4.310441	4.887112
15	3.642482	4.177248	4.784589	5.473566
16	3.970306	4.594973	5.310894	6.130394
17	4.717120	5.559917	6.543553	7.689966
20	5.604411	6.727500	8.062312	9.646293
22	6.658600	8.140275	9.933574	12.100310
24	7.911083	9.849733	12.239157	15.178629
25	8.623081	10.834706	13.585464	17.000064
26	9.399158	11.918177	15.079865	19.040072
28	11.167141	14.420994	18.579901	23.883866
30	13.267678	17.449402	22.892297	29.959922
32	15.763329	21.113777	28.205599	37.581726
34	18.728411	25.547670	34.752118	47.142517
35	20.413968	28.102437	38.574851	52.799962
40	31.409420	45.259256	65.000867	93.050970
45	48.327286	72.890484	109.53024	163.98761
50	74.357520	117.39085	184.56482	289.00219
55	114.40826	189.05914	311.00246	509.32061
60	176.03129	304.48164	524.05724	897.59693

Table A

Amount of 1 at Compound Interest

COMPOUNDED
YEARLY, TWICE YEARLY, QUARTERLY
(continued)

N	1.13	1.14	1.15	1.16
2	1.2769	1.2996	1.3225	1.3456
3	1.442897	1.481544	1.520875	1.560896
4	1.630474	1.68896	1.749006	1.810639
5	1.842435	1.925415	2.011357	2.100342
6	2.081952	2.194973	2.313061	2.436396
7	2.352605	2.502269	2.660020	2.826220
8	2.658444	2.852586	3.059023	3.278415
9	3.004042	3.251949	3.517876	3.802961
10	3.394567	3.707221	4.045558	4.411435
12	4.334523	4.817905	5.350250	5.936027
14	5.534753	6.261349	7.075706	7.987518
15	6.254270	7.137938	8.137062	9.265521
16	7.067326	8.137249	9.357621	10.748004
18	9.024268	10.575169	12.375454	14.462514
20	11.523088	13.743490	16.366537	19.460759
22	14.713831	17.861039	21.644746	26.186398
24	18.788091	23.212207	28.625176	35.236417
25	21.230542	26.461916	32.918953	40.874244
26	23.990513	30.166584	37.856796	47.414123
28	30.633479	39.204493	50.065612	63.800444
30	39.115898	50.95159	66.21772	85.849877
32	49.947090	66.214826	87.565068	115.51959
34	63.777439	86.052788	115.804803	155.44316
35	72.068506	98.100178	133.175523	180.31407
40	132.781552	188.883514	267.863546	378.72115
45	244.641402	363.679072	538.769269	795.44382
50	450.735925	700.23988	1083.65744	1670.7038
55	830.451725	1348.23880	2179.62218	3509.0488
60	1530.05347	2595.91866	4383.99874	7370.2014

Table A
Amount of 1 at Compound Interest

COMPOUNDED
YEARLY, TWICE YEARLY, QUARTERLY
(continued)

N	1.17	1.18	1.19	1.20
2	1.368900	1.392400	1.416100	1.440000
3	1.601613	1.643032	1.685159	1.72800
4	1.873887	1.938778	2.005339	2.073600
5	2.192448	2.287758	2.386354	2.488320
6	2.565164	2.699554	2.839761	2.985984
7	3.001242	3.185474	3.379315	3.583181
8	3.511453	3.758859	4.021385	4.299817
9	4.108400	4.435454	4.785449	5.159780
10	4.806828	5.233836	5.694684	6.191736
12	6.580067	7.287593	8.064242	8.916100
14	9.007454	10.147244	11.419773	12.839185
15	10.538721	11.973748	13.589530	15.407022
16	12.330304	14.129023	16.171840	18.488426
18	16.878953	19.673251	22.900518	26.62333
20	23.105599	27.393035	32.429423	38.337600
22	31.629255	38.142061	45.923307	55.206144
24	43.297287	53.109006	65.031994	79.496847
25	50.657826	62.668627	77.388073	95.396217
26	59.269656	73.948980	92.091807	117.47546
28	81.134232	102.966560	130.41208	164.84466
30	111.064650	143.370638	184.675312	237.37631
32	152.036399	199.629277	261.518710	341.82189
34	208.122627	277.963805	370.336645	492.22352
35	243.503474	327.997290	440.700607	590.66823
40	533.868713	750.378345	1051.66751	1469.7715
45	1170.47941	1716.68388	2509.65060	3657.2619
50	2566.21528	3927.35686	5988.913902	9100.4382
55	5626.29366	8984.84112	14291.6666	22644.802
60	12335.3565	20555.1399	34104.9709	56347.514

Amount of 1 at Compound Interest

COMPOUNDED
YEARLY, TWICE YEARLY, QUARTERLY
(continued)

N	1.21	1.22	1.23	1.24
2	1.464100	1.488400	1.512900	1.539600
3	1.771561	1.815848	1.860867	1.906624
4	2.143589	2.215335	2.288866	2.364214
5	2.593742	2.702708	2.815306	2.931625
6	3.138428	3.297304	3.462826	3.635215
7	3.797498	4.022711	4.289276	4.507667
8	4.594973	4.907707	5.238909	5.589507
9	5.559917	5.3987403	6.443859	6.930988
10	6.727500	7.304631	7.925946	8.594426
12	9.849733	10872213	11.991164	13.214789
14	14.420994	16.182202	18.141432	20.319059
15	17.449402	19.742287	22.313961	25.195633
16	21.113777	24.085590	27.446172	31.242585
18	30.912681	35.848992	41.523314	48.038599
20	45.259256	53.357640	62.820622	73.864150
22	66.264076	79.417512	95.041318	113.57351
24	97.017234	118.20502	143.78801	174.63063
25	117.39085	144.21013	176.85925	216.54199
26	142.04932	175.93636	329.11155	41286416
28	204.96506	261.86368	329.11155	412.86416
30	304.48164	389.75789	497.91286	634.81993
32	445.79157	580.11565	753.29237	976.09912
34	652.68344	863.44413	1139.6560	1555.8500
35	789.74696	1053.4018	1401.7769	1861.0540
40	2048.4002	2847.0378	3946.4305	5455.9126
45	5313.0226	7694.7122	11110.408	15994.690
50	13780.612	20796.562	31279.195	46890.434
55	35743.359	56207.036	88060.496	137465.17
60	92709.069	151911.22	247917.21	402996.34

Table B
Amount of 1 at Compound Interest

COMPOUNDED MONTHLY

$$\left(1 + \frac{\text{Interest Rate}}{\text{Number of compound periods in 1 year}}\right)^{N}$$

N = number of compound periods during program

Example - 6 % interest compounded monthly for 5 years

$$\left(1 + \frac{.06}{12}\right)^{(12 \times 5)} = \left(1 + .005\right)^{60} = \left(1.005\right)^{60} = 1.3488502$$

N	$1 + \frac{05}{12}$ 1.0041667	$1 + \frac{.06}{12}$ 1.005	$1 + \frac{.07}{12}$ 1.0058333	$1 + \frac{.08}{12}$ 1.0066667
2	1.0083508	1.010025	1.0117006	1.0133778
3	1.0125523	1.0150751	1.0176022	1.0201346
4	1.0167713	1.0201505	1.0235382	1.0769347
5	1.0210078	1.0252513	1.0295088	1.0337809
6	1.0252621	1.0303775	1.0355142	1.0406728
10	1.042457	1.0511401	1.0598883	1.068703
12	1.0511623	1.0616778	1.0722897	1.0829999
18	1.0777169	1.0939389	1.1103719	1.1270486
24	1.1049422	1.1271598	1.1498051	1.1728889
30	1.1328554	1.1614001	1.1906289	1.2205936
36	1.1614736	1.1966805	1.2329241	1.2702386
42	1.1908149	1.2330327	1.2767104	1.3219028
48	1.2208973	1.2704892	1.3220518	1.3756683
60	1.2833612	1.3488502	1.4176224	1.4898487
61	1.2887060	1.3555944	1.4258919	1.4997810
72	1.3490210	1.4320443	1.5201019	1.6135060
96	1.4905902	1.6141427	1.7478209	1.8924632
120	1.6470161	1.8193967	2.0096534	2.2196491
180	2.1137166	2.4540936	2.8489297	3.3069412
240	2.7126619	3.3102045	2.0387067	4.9268419
300	3.4813251	4.4649698	5.7253613	7.3402489
360	4.4677977	6.0225752	8.1164974	10.93586

M O N T H S

Table B
Amount of 1 at Compound Interest

COMPOUNDED MONTHLY
(continued)

N	$1 + \dfrac{.09}{12}$ 1.0075	$1 + \dfrac{.10}{12}$ 1.008333	$1 + \dfrac{.11}{12}$ 1.0091667	$1 + \dfrac{.12}{12}$ 1.01
2	1.015056	1.016736	1.0184174	1.0201
3	1.0226692	1.0252088	1.0277529	1.030301
4	1.0303392	1.0337522	1.0371739	1.040604
5	1.0380667	1.0423668	1.0466814	1.0510101
6	1.0458522	1.0510531	1.0562759	1.0615202
10	1.0775826	1.0865284	1.0955419	1.1046221
12	1.0938069	1.1047087	1.1157188	1.1269825
18	1.1439604	1.1611116	1.1785077	1.1961475
24	1.1964135	1.2203813	1.2448285	1.2697347
30	1.2512718	1.2826947	1.3148824	1.3478489
36	1.3086454	1.3481802	1.3888786	1.4307688
42	1.3686497	1.4169913	1.4670391	1.5187899
48	1.4314053	1.4893517	1.5495981	1.6122261
60	1.5656810	1.6453057	1.7289157	1.8166967
72	1.7125527	1.8175943	1.9289839	2.0470993
96	2.0489212	2.2181686	2.4012541	2.5992729
120	2.4513571	2.7070308	2.9891496	3.3003869
180	3.8380433	4.4538931	5.1679878	5.9958020
240	6.0091515	7.3280155	8.9350154	10.832554
300	9.4084145	12.056783	15.447889	19.788466
360	14.730576	19.837163	26.708089	35.949641

M O N T H S

Table B
Amount of 1 at Compound Interest

COMPOUNDED MONTHLY
(continued)

MONTHS N	$1 + \cdot\frac{13}{12}$	$1 + \cdot\frac{14}{12}$	$1 + \cdot\frac{15}{12}$	$1 + \cdot\frac{16}{12}$
	1.0108333	1.0116667	1.0125	1.0133333
2	1.021784	1.0234694	1.0251563	1.0268444
3	1.0328533	1.0354099	1.0379707	1.0405356
4	1.0440425	1.0474897	1.0509453	1.0544094
5	1.0553529	1.0597104	1.0640822	1.0684681
6	1.0667858	1.0720737	1.0773832	1.0827143
10	1.1137697	1.1229865	1.1322708	1.1416245
12	1.1380320	1.1493425	1.1607545	1.1722708
18	1.2140365	1.2321801	1.2505774	1.2692347
24	1.2951169	1.3209881	1.3473511	1.3742188
30	1.3816124	1.4161969	1.4516134	1.4878867
36	1.4738845	1.5182678	1.5639438	1.6109566
42	1.5723191	1.6276953	1.6849668	1.7442062
48	1.6773278	1.7450097	1.8153549	1.8894774
60	1.9088528	2.0056138	2.1071814	2.2138069
72	2.1723356	2.3051371	2.4459203	2.5951812
96	2.8134285	3.0450588	3.2955132	3.5663355
120	3.6437189	4.0224866	4.4402132	4.9009409
180	6.9553228	8.0675544	9.3563345	10.849737
240	13.276687	16.0180398	19.715494	24.019222
300	25.343241	32.451308	41.544120	53.173394
360	48.376515	65.084662	87.540995	117.71539

Table B

Amount of 1 at Compound Interest

COMPOUNDED MONTHLY

(continued)

N	$1 + \frac{.17}{12}$ 1.0141667	$1 + \frac{.18}{12}$ 1.015	$1 + \frac{.19}{12}$ 1.0158333	$1 + \frac{.20}{12}$ 1.0166667
2	1.028534	1.030225	1.0319173	1.033611
3	1.0431049	1.0456784	1.0482559	1.050838
4	1.0578823	1.0613636	1.0648533	1.0683519
5	1.0728689	1.077284	1.0817134	1.086158
6	1.0880679	1.0934433	1.0988405	1.1042606
10	1.1510477	1.1605408	1.170104	1.1797391
12	1.1838917	1.1956182	1.2074505	1.2193916
18	1.2881546	1.3073406	1.3267956	1.3465261
24	1.4015996	1.4295028	1.4579368	1.4869158
30	1.5250355	1.5630802	1.6020400	1.6419426
36	1.6593422	1.7091395	1.7603865	1.8131326
42	1.8054795	1.8688471	1.9343841	2.0021709
48	1.9644815	2.0434783	2.1255796	2.2109186
60	2.3257334	2.4432198	2.5665322	2.6959754
72	2.7534166	2.921158	3.0989607	3.2874497
96	3.8591998	4.1758035	4.5180887	4.8881609
120	5.4090359	5.9693229	6.5870876	7.2682836
180	12.579976	14.584368	16.905973	19.595114
240	29.257669	36.632816	43.389723	52.827946
300	68.045539	87.058800	111.36112	142.42285
360	158.25578	212.70380	285.81191	383.9685

MONTHS

When you examine the above table as well as the previous 3 pages you can tell how long it takes to double, triple or for that mater even increase 10 fold. In the above chart it takes about 48 months to double your money at 18 % interest but only takes 42 months at 20 %. Likes wise slightly less than 180 months to equal 10 times more at 17 %.

Table C
Amount of 1 at Compound Interest

COMPOUNDED DAILY

$$\left(1 + \frac{\text{Interest Rate}}{\text{Number of compound periods in 1 year}}\right)^{\textbf{N}}$$

N = number of compound periods during program

Example - 7 % interest compounded daily for 6 years

$$\left(1 + \frac{.07}{365}\right)^{365 \times 6} = \left(1 + .00019178\right)^{2190} = \left(1..0001918\right)^{2190} = 1.5219002$$

N	$1 + \frac{.05}{365}$	$1 + \frac{.06}{365}$	$1 + \frac{.07}{365}$	$1 + \frac{.08}{365}$
	1.000137	1.0001644	1.0001918	1.0002192
Days				
90	1.0124042	1.0149033	1.0174084	1.0199197
180	1.0249623	1.0300287	1.0351199	1.0402361
365	1.0512675	1.0618313	1.0725085	1.0832776
730	1.1051634	1.1274858	1.1502583	1.1734903
1095	1.1618223	1.1971997	1.2336791	1.2712157
1460	1.2214105	1.2712242	1.3230943	1.3770795
1825	1.2840034	1.3498256	1.4190199	1.4917594
2190	1.3498311	1.4332871	1.5219002	1.6129895
2920	1.4917838	1.6160108	1.7505784	1.8964692
3650	1.6486648	1.8220292	2.0136174	2.2255238
4380	1.8220439	2.0544595	2.3161802	2.6116724
5110	2.0136562	2.3162075	2.6642056	3.0644784
5840	2.2255970	2.6117411	3.0645247	3.5961357
7300	2.7180956	3.3197304	4.0552233	4.9521648
9125	3.4904804	4.4811382	5.7546440	7.3874382
10950	4.4812286	6.0487552	8.1645239	11.020228
12775	5.7549199	8.1664752	11.588463	16.444205
14600	7.3895216	11.023647	16.444836	24.531777

Table C

Amount of 1 at Compound Interest

COMPOUNDED DAILY
(continued)

N Days	$1 + \dfrac{.09}{365}$ 1.0002466	$1 + \dfrac{.10}{365}$ 1.000274	$1 + \dfrac{.11}{365}$ 1.0003014	$1 + \dfrac{.12}{365}$ 1.0003288
90	1.0224371	1.0249606	1.0274903	1.0300261
180	1.0453775	1.0505442	1.0557363	1.0609538
365	1.0941621	1.1051558	1.1162596	1.1274746
730	1.1971908	1.2213693	1.2460355	1.2712295
1095	1.3099652	1.3498033	1.3909449	1.4332446
1460	1.4332657	1.4918026	1.5526044	1.6159469
1825	1.5682251	1.6486084	1.7331095	1.8219391
2190	1.7158925	1.8219691	1.9346001	2.0543380
2920	2.0542506	2.2252971	2.4105804	2.6112843
3650	2.4593299	2.7179095	3.0036686	3.3194619
4380	2.9446055	3.3195713	3.7426776	4.2196966
5110	3.5248732	4.0544224	4.6635091	5.3640741
5840	4.2199456	4.9519471	5.8108977	6.8188056
7300	6.0403034	7.3870322	9.0220252	11.018827
9125	9.4851001	12.178323	15.636158	20.075632
10950	14.874773	20.077285	27.099174	36.576578
12775	23.326993	33.142934	46.965836	66.640296
14600	36.581974	54.568245	81.396938	121.41456

Table C
Amount of 1 at Compound Interest

COMPOUNDED DAILY
(continued)

N Days	$1 + \frac{.13}{365}$ 1.0003562	$1 + \frac{.14}{365}$ 1.0003836	$1 + \frac{.15}{365}$ 1.000411	$1 + \frac{.16}{365}$ 1.0004384
90	1.0325715	1.0351200	1.0376748	1.0402357
180	1.0662039	1.0714735	1.0767689	1.0820904
365	1.1388168	1.150259	1.1618159	1.1734885
730	1.2969038	1.3230958	1.3498161	1.3770753
1095	1.4769358	1.5219029	1.5682378	1.615982
1460	1.6819594	1.7505825	1.2080665	1.8963363
1825	1.9154436	2.0136234	1.4035508	2.2253288
2190	2.1813394	2.3161884	2.4593697	2.6113977
2920	2.8289873	3.0645392	3.3196968	3.5960912
3650	3.6689243	4.054679	4.4809801	4.9520882
4380	4.7582417	5.3647288	6.0484991	6.8193981
5110	6.1709815	7.0980502	8.1643615	9.3908244
5840	8.0031691	9.3914005	11.020387	12.931872
7300	13.461005	16.440422	20.079183	24.523178
9125	25.783797	33.104817	42.504269	54.572133
10950	49.387409	66.660633	89.097442	121.44094
12775	94.598797	134.22941	190.46078	270.24602
14600	181.19866	270.28747	403.17359	601.38624

Tax Brackets -

Form 1040, line 39		Single Taxpayer (2001 tax table)			of the amount over
Over	but not over	basic tax	percent more		
$ 0	27,050	--------	15%		$0
27,050	65,550	4,057.50	+	27.5%	27,050
65,550	136,750	14,645.00	+	30.5%	65,550
136,750	297,350	36,361.00	+	35.5	136,750
297,350	---------	93,374.00	+	39.1%	297,350

If a person makes $27,050 the tax rate is 15%
$27,050 X .15 = $4,057.50

For someone making $50,000 the amount of tax would be-

(the basic tax) $ 4,057.50

You then determine the amount over 27,050

50,000 minus 27,050 equaling $ 22,950
The $22,950 is then taxed at 27.5% equaling $ 6,311.25

6,311.25 is then added to the basic tax of 4,057.50 =
$ 10,368.75

To find the percent of income going towards tax divide

10,368.75 by 50,000 = 0.207375 rounded of to 20.7 %

Going to the Next Higher Tax Bracket

Example - Taxpayer A making $136,700 will pay-

14,645.00 (the basic tax) plus (136,700 minus 65,550
equaling 71,150.) $ 71,150 is then taxed at 30.5%

71,150 X 0.305 = $21,700.75
basic tax   *over amount*
14,645.00 + 21,700.75 = $36,345.75

Going to the Next Higher Tax Bracket (continued)

Taxpayer B making $ 136,800 ($100 more than A) will pay

$36,361 plus 35.5 % of the amount over $136,750.00

> 136,800 - 136,750 = $ 50.00
> $ 50.00 X 0.355 = 17.75 to be added to the basic tax
> 36,361 + 17.75 = $ 36,378.75
>
> Compare the two tax payers
> A made $ 136,700 and pays $ 36,345.75
> B made $ 136,800 and pays $ 36,378.75

Tax payer B made $100.00 more than A and paid $ 33.00 more
in tax. The amount in additional tax is proportional to the
additional money made. Both paid about the same percentage
even though each were in different tax brackets.

> A 36,345.75 divided by 136,700 = 0.26588
> B 36,378.75 divided by 136,800 = 0.26593

A quick way to determine your increased taxes, when making
more money, is to look at your tax bracket percent and multiply
it times the additional money you will be making.

> **Example-** Taxpayer C is making $ 65,550 and will pay
> $ 14,645.00 in income taxes. For every dollar
> over $ 65,550 they will pay 30.5 cents.

The tax brackets are based on the last dollars you make
and not the first dollars.

> **Example -** A tax payer making $297,350 would pay
> $93,374.00 in income tax. This is 31.4 % of
> their income. However for every dollar they
> make over 297,350 they will pay the rate of
> 39.1%
>
> 93,374 divided by 297,350 = .314 = 31.4 %

Going to the Next Higher Tax Bracket (continued)

Example- Another tax payer making $2,000,000 will pay an amount that is closer to 39.1 %.

The basic tax is $93,374.00 . The percent more tax is
2,000,000 - 297,350 = 1,702,650 X .391= $665,736.15

93,374.00 + 665,736.15 = 759,110.15 total tax

759,110.15 divided by 2,000,000 = 0.3795 = 37.9 %

The more money a taxpayer makes the closer they get to the maximum 39.1% rate. After they have made $297,350 and paid $93,374 in income tax, for every dollar they make, they will have to pay 39.1 cents.

Estimating Tax Savings of a Tax Write Off

When you encounter a tax write off, it decreases your income, thus the amount of tax you pay. To determine how much you will save in taxes, multiply the cost of the item, times the tax bracket you are in.

Example- In 2001 if you made $52,032 , your Federal Tax Rate for the last dollars you made is 27.5%. If you encounter a tax write off of $1,500, how much will you save in taxes?

1,500 X .275 = 420

Because you are paying less tax money, you could consider that the item cost you less than the purchased price.

$1,500 minus $412.50 = $1087.50

Taxes

Penalty and Interest

Most tax authorities such as sales tax and employee tax boards charge a penalty and interest for late payments. Most taxing authorities charge a penalty if you file after the due date even if you send the payment with the filing.

Example- A business can not pay the $ 2,500.00 they owe for the collection of sales tax they have collected on sales. The penalty is 10% for paying late and 1% per-month or a portion of a month for interest. How much do they owe if payment is made within the next month?

Amount owed +(amount owed X 0.10) + (amount owe X 0.01)

$$\text{2,500} + \underset{\textit{late payment penalty}}{(\text{2,500 X 0.10})} + \underset{\textit{interest}}{(\text{2,500 X 0.01})} =$$

2,500 + 250.00 + 25.00 = $ 2,775.00

or do it the fast way
amount owed X (1 + late penalty + interest)
2,500   X (1 + 0 .10 + 0.01)
2,500   X   1.11 = $ 2,775.00

Some tax authorities do not compound interest. In these few instances the interest is simply multiplied times the number of months late.

Example - John owes $ 940.00 for sales tax he has collected though sales but has not paid for 3 1/2 months. They charge 10 % penalty and 1 1/4 % per-month or portion of a month. How much does he owe?

940 + (940 X 0.10) + (940 X (4 X .0125) =
940 + (940 X 0.10) + (940 X 0.05) =

940 + 94 + 47.00 = $ 1,081.00

Penalty and Interest (continued)

A faster way to do it is to find just the total

940.00 X 1.15 = $ 1,081.00

Compounding the Interest

Using the previous example- $ 940, 3 1/2 months, 10% late penalty and 1 1/4 % interest each month. Please note , in most cases the interest is compounded but the late payment penalty is paid only once.

If it was 1 month late-

940 + (940 X 0.10) + (940 X 0.0125)

940 + 94 + 11.75 = 1045.75

If it was 3 1/2 months late-

number of months
940 X { 10 % + (1 + interest) }

$$940 \text{ X } \{0.10 + (1 + 0.0125)^4 \}$$

using a calculator do the following
1.0125 $\boxed{Y^x \text{key}}$ 4 $\boxed{=}$ 1.0509453

940 X (0.10 + 1.0509453)

940 X 1.1509453 = $ 1,081.89 is due after 3 1/2 months

When you use the Y^x key you are doing the following

1.0125 X 1.0125 X 1.0125 X 1.0125 = 1.0509453

Please see page 9:18 to further explain compounding interest.

Penalties On Income Taxes

If you file on time and do not pay any or only a
portion of the amount due, the following applies
to the unpaid amount.
1. Late payment penalty is 1/2 of 1 percent (0.005)
for each month or portion of a month.
It increases to 1 percent after IRS issues a
"Notice of Intent to Levy" or files a lien.
The maximum the IRS can charge for late
payment penalty is 25 percent.
2. The interest rate is based on the Federal Short Term
Rate plus 3 percent. It can change quarterly.
Presently it is 4 % Fed plus 3 % equaling an
annual percentage rate of 7 %.
To find the daily interest divide the rate by 365.
0.07 / 365 = 0.0001918
It is compounded daily and applies to the unpaid
tax plus the monthly penalty.

If you do not file an Income Tax Return or if you file
it late the following applies to the time period
that it was late.
1. The late filing penalty is 4 1/2 percent and
increases 1/2 percent per month up to 25%.
2. The late payment penalty is 1/2 of one percent.
and increases 1/2 percent per month up to 25%.
3. The interest rate is the same as above. It applies to
the amount not paid and the penalties that are
incurred.

If you have an installment agreement the following
applies.
1. The installment penalty is 1/4 of 1 percent per month
(0.0025)
2. You still get charged the same interest as above,
presently an annual percentage rate of 7 percent
applied against the unpaid tax plus the penalty
applied each month

Please refer to the following examples

Example- Filed on time but did not pay the amount due of
$ 3,000 for **4 months.**

Late Payment Penalty

$ 3,000 X (0.005 X 4 months)

$ 3,000 X 0.02 = $ 60.00

Interest applied to the amount owed plus the
late payment penalty.

compounding interest is explained on page 9:18

$$\$\,3{,}060 \; X \; \left(1 + \frac{0.07}{365}\right)^{\frac{4 \text{ months } X\, 30}{120}}$$

.07 divided by 365 = 0.0001918

$ 3,060 X (1.0001918)

Do the following on a calculator

$.07 \div 365 = .0001918 + 1 \;\boxed{=}\; \boxed{y^x \text{key}}\; 120 = 1.0232783$

$ 3,060 X 1.0232783 = $ 3,131.23 includes
the amount owed plus the late payment penalty
and interest.

Example- Filed on time but did not pay the amount due of
$ 3,000 for **20 months.** After 12 months the
IRS files an "Intend to Levy."

Late Payment Penalty

$ 3,000 X (0.005 X 12 months)

$ 3,000 X 0.06 = $ 180.00

After 12 months the IRS files an "Intend to
Levy," the penalty increases to 0.01 per month

$ 3,000 X (0.01 X 8 months)

$ 3,000 X 0.08 = $ 240.00

(*continued*)

Taxes IRS

Total late payment penalty
$$\$\,180.00 + 240.00 \;=\; \$\,420.00$$

Interest applied to the amount owed plus the penalty.

20 months X 30 days
$$3{,}420 \;\; X \;\; 1 +\!\left(\frac{0.07}{365}\right)^{\!600}$$
$$3{,}420 \;\; X\,(1.0001918\,)$$

Do the following on a calculator

$.07\;\boxed{\div}\;365\;\boxed{=}\;.0001918\;+\;1\;=\;\boxed{y^{x}key}\;600\;\boxed{=}\;1.1219379$

$$3{,}420 \;\; X \;\; 1.1219379 \;\; = \$\,3{,}837.03 \text{ includes the}$$
penalty and the interest

Example- Did <u>not</u> file on time or pay the amount due of
$\$\,3{,}000$ for **4 months.**

Late Filing Penalty
$$\$\,3{,}000 \;\; X \;\; \underset{\text{1st month}}{0.045} \;+\; (\; \underset{\text{3 months}}{0.005 \;\; X \;\; 3 \text{ months}} \;)$$
$$\$\,3{,}000 \;\; X \;\; 0.045 \;+\; 0.015 \;=$$

$$\$\,3{,}000 \;\; X \;\; 0.06 \;=\; \$\,180.00$$

Late Payment Penalty
$$\$\,3{,}000 \;\; X \;\; (\,0.005 \;\; X \;\; 4 \text{ months}\,)$$

$$\$\,3{,}000 \;\; X \;\; 0.02 \;=\; \$\,60.00$$
Total penalties $\$\,180.00 + 60.00 \;=\; \$\,240.00$

Interest applied to the amount owed plus the penalties.

4 months X 30
$$\$\,3{,}240 \;\; X \;\; \left(\,1 +\frac{0.07}{365}\,\right)^{\!120}$$
$$\$\,3{,}240 \;\; X \;\; 1.0001918\,)$$
calculate $1.0001918\,\boxed{Y^{x}\,key}\;\boxed{=}\;1.0232783$
$$\$\,3{,}240 \;\; X \;\; 1.0232783 \;=\; \$\,3{,}315.42$$

Example - Did not file on time or pay the amount due of $ 3,000 for **20 months.** After 12 months the IRS files an "Intend to Levy."

Late filing Penalty

$$\$\, 3,000 \ \text{X} \ \underset{\text{1st month}}{0.045} + \ (0.005 \ \text{X} \ 19 \ ) \ =$$

$$\$\, 3,000 \ \text{X} \ 0.045 \ + \ 0.095 \ =$$

$$\$\, 3,000 \ \text{X} \ 0.14 \ = \ 420$$

Late Payment Penalty

$$\$\, 3,000 \ \text{X} \ (\ 0.005 \ \text{X} \ 12 \ ) \ =$$

$$\$\, 3,000 \ \text{X} \ 0.06 \ = \ \$\, 180.00 \ \text{for 12 months}$$

8 months at 0.01 because IRS filed and "Intend to Levy."

$$\$\, 3,000 \ \text{X} \ 0.08 \ + \$\, 240.00$$

Total late payment penalty $ 180 + 240 = $ 420.00

Total penalties for late filing and late payment =

$$\$\, 420.00 + \ 420.00 \ = \$\, 840.00$$

Interest applied to the amount owed and the penalties

$$\$\, 3,840.00 \ \text{X} \ 1 + \left(\frac{0.07}{365} \ \right)^{20 \text{ months X } 30}$$

$$\$\, 3,840.00 \ \text{x} \ (\ 1.0001918 \ )^{600}$$

Calculate 1.0001918 Y^x key 600 = 1.1219508

$$\$\, 3,840 \ \text{X} \ 1.1219508 \ = \ \$\, 4,325.12$$

Property Tax

Property taxes vary greatly from state to state, as to what is taxed and how the tax is figured. Some state the rate as per $1,000; as per $100 and as a mill per an amount (1 mill = 1/1000 of $1.00 or 1,000 mills = $1.00). Some states use the fair market value to base the tax, while some use an assessed value, using an assessment ratio. The following is an example of the 3 different ways the tax rate is stated.

Percent of the Value or Assessed Value

This is the easiest to figure and understand.Usually it is stated as a percent of the value of the real estate, minus exceptions, such as $ 7,000 for living in the home that is taxed.   Added to this rate is for local special assessments like sewage, library, special parks, or extras a community votes for or allows.

Value of the home minus exception X (0.01 + 0.00067) = tax
county   local

$ 380,000 - 7,000 X 0.01067 = $ 3,979.91 tax

Assessment per $100 or $1,000

This is usually sated as dollar amount per $1,000 or $100 of the assessed value

$ 3.20 per $100.00

The tax for a home assessed for $62,500.00
62,500 divided by 100 = 625   625 X 3.20 = $ 2,000.00

$ 28.50 per $ 1,000.00

The tax for a home assessed for $75,000.00
75,000 divided by 1,000 = 75   75 X 28.50 = $ 2,137.50

Mils

Usually stated as so many mils. Meaning for every $1,000 value the tax was per mils.

33 mil tax on a home valued at $58,000.00
33 X 58 = 1,914.00

Social Security

As of the year 2002, the Social Security Tax on the amount of money you make is 7.65 percent. This is made up as,

You the <u>employee</u> pays 6.2 % for Retirement & Disability Insurance (OASDI) and 1.45 % for Health Insurance

Retirement and Disability Insurance is broken down as; 5.35 % for Retirement and 0.85 % for Disability Insurance = 6.2 %

Your <u>employer</u> pays on your account the same amount as you do.

Retirement Old-Age and Survivors Disability Insurance (OASDI)
Presently you are taxed for retirement on the first $84,900 you make. In other words if you make $42,000, you will be taxed for the full $42,000. (The limit of $84,900 will go up in the future every year to keep up with average wage levels.)

$$\$42,000 \text{ X } .062 = \$2,604.00$$

If you make $98,100, you will only be taxed on $84,900

$$\$84,900 \text{ X } .062 = \$5,263.80$$

Health Insurance (HI)
For Health Insurance (Medicare) you will be taxed on every cent you make at 1.45 percent (0.0145).

$$\$42,000 \text{ X } 0.0145 = \$609$$
Or if you make
$$\$250,000 \text{ X } .0145 = \$3,625.00$$

Self-Employed
If you are self-employed you pay the amount an employer pays plus the amount an employee pays. A total of 15.3 %.

12.4% Retirement Disability Insurance
2.9% Health Insurance (Medicare)

Estimating Social Security Retirement Benefit
for Workers Born in 1940

Please refer to the following page

Step 1

> Enter how much you made for each year in Column B. If you made less than the amount listed in Column A, put that amount in Column B. If you made more than what is listed in Column A, put the amount listed in Column A in Column B.

Step 2

> Multiply the amounts you listed in Column B by the index factors in Column C, and enter the results in Column D. The figure you entered in Column D is what the earnings you listed in Column B, would be worth today.

Step 3

> Choose from Column D the 35 years with the highest amounts.

> Enter here $_____

Step 4

> Divide the result from Step 3 by 420 (this is the number of months in 35 years).

> Enter here $_____

Step 5

> a- Multiply the first $592 in Step 4 by 90%.

> Enter here $_____

> b- Multiply any amount in Step 4 over $592 and less than or equal to $3,567 by 32 %. (0.32)

> Enter here $_____

> c- Multiply any amount over $3,567 in Step 4 by 15% (0.15).

> Enter here $_____

Step 6

> Add a, b and c from Step 5. This is your estimated monthly retirement benefit at age 65 and 4 months.
> *(your full retirement age)*

> Total benefit $_____

Step 7

> If you are retiring at age 62 multiply Step 6 by 77.5 %

For Workers Born in 1940
*(please note that the following earnings and index factors change,
as to the year you were born)*

Year	A. Maximum Earnings	B. Actual Factor	C. Index	D. Indexed Earnings
1951	3,600		11.49	
1952	3,600		10.81	
1953	3,600		10.24	
1954	3,600		10.19	
1955	4,200		9.74	
1956	4,200		9.10	
1957	4,200		8.83	
1958	4,200		8.75	
1959	4,800		8.34	
1960	4,800		8.02	
1961	4,800		7.87	
1962	4,800		7.49	
1963	4,800		7.31	
1964	4,800		7.03	
1965	4,800		8.90	
1966	6,600		6.51	
1067	6,600		6.17	
1968	7,800		5.77	
1969	7,800		5.46	
1970	7,800		5.20	
1971	7,800		4.95	
1972	9,000		4.51	
1973	10,800		4.24	
1974	13,200		4.00	
1975	14,100		3.73	
1976	15,300		3.49	
1977	16,500		3.29	
1978	17,700		3.05	
1979	22,900		2.80	
1980	25,900		2.57	
1981	29,700		2.33	
1982	32,400		2.21	
1983	35,700		2.11	
1984	37,800		1.99	
1985	39,600		1.91	
1986	42,000		1.86	
1987	43,800		1.75	
1988	45,000		1.66	
1989	48,000		1.60	
1990	51,300		1.53	
1991	53,400		1.47	
1992	55,500		1.40	
1993	57,600		1.39	
1994	60,000		1.35	
1995	61,200		1.30	
1996	62,700		1.24	
1997	65,400		1.17	
1998	98,400		1.11	
1999	72,600		1.06	
2000	76,200		1.00	
2001	80,400		1.00	

11 : 14 Estimating Social Security Retirement Benefit
For Workers Born in 1940
*(please note that the following base amounts and percentages change,
as to the year you were born)*

Example

Please refer to the example page of John Doe. In 2001 John is 61 and is going to estimate his retirement benefit. He has worked since he was 18 and has worked 43 years, starting in 1958

Step 1   John Doe entered his actual earnings in Column B, but not more than the amount shown in Column A. . Please note that some of the years he worked he made more than the maximum earnings, however only the amount that he is taxed on, is listed.

Step 2

Each of the amounts in Column B. were multiplied by the Index Factors in Column C and listed in Column D.

Step 3

The total of his highest 35 years of his Indexed earnings is $ 2,250,215.00

Step 4   *12 months X 35 years = 420*
$ 2,250,215 is divided by 420 equaling $ 5,358

Step 5

a. $ 592 is multiplied by 90% (.90) equaling $ 533
b. $ 3567 is multiplied by 32% (.32) equaling $ 1141
c. Multiply 5,358 by 15% (.15) equaling $ 804

Step 6

Add the totals from Step 5 a,b and c, equaling
$ 2,478.00

The amount John Doe should get each month for retirement is $ 2,478.00 However the maximum benefit for 2002 is $ 1,660.00 per month. Social Security benefits are increased every year. For the last few years it has increased an average of 2.9 % over each previous year. This would then give John Doe $ 1889.00 per month. He is presently 61 and full benefits are at 65 1/2 years.
Calculate 1.029 Y^x Key 4.5 X 1660 = $1,887.89

Year	A. Maximum Earnings	B. Actual Earnings	C. Index Factor	D. Indexed Earnings
1951	3,600		11.49	
1952	3,600		10.81	
1953	3,600		10.24	
1954	3,600		10.19	
1955	4,200		9.74	
1956	4,200		9.10	
1957	4,200		8.83	
1958	4,200	3,800	8.75	33,250
1959	4,800	4,750	8.34	39,615
1960	4,800	4,800	8.02	38,496
1961	4,800	4,800	7.87	37,776
1962	4,800	4,800	7.49	35,952
1963	4,800	4,800	7.31	35,082
1964	4,800	4,800	7.03	33,744
1965	4,800	4,800	6.90	33,120
1966	6,600	5,890	6.51	38,344
1967	6,600	6,600	6.17	40,722
1968	7,800	7,800	5.77	45,006
1969	7,800	7,800	5.46	42,588
1970	7,800	7,800	5.20	40,560
1971	7,800	7,800	4.95	38,610
1972	9,000	9,000	4.51	40,590
1973	10,800	10,700	4.24	45,368
1974	13,200	12,700	4.00	50,800
1975	14,100	13,975	3.73	52,127
1976	15,300	14,890	3.49	51,966
1977	16,500	16,500	3.29	54,285
1978	17,700	16,800	3.05	51,240
1979	22,900	18,975	2.80	53,130
1980	25,900	24,780	2.57	63,685
1981	29,700	28,750	2.33	66,988
1982	32,400	16,500	2.21	36,465
1983	35,700	34,750	2.11	73,323
1984	37,800	37,800	1.99	75,222
1985	39,600	39,030	1.91	74,547
1986	42,000	42,000	1.86	78,120
1987	43,800	43,600	1.75	76,300
1988	45,000	45,000	1.66	74,700
1989	48,000	47,200	1.60	75,520
1990	51,300	51,200	1.53	78,336
1991	53,400	53,400	1.47	78,498
1992	55,500	54,000	1.40	75,600
1993	57,600	57,300	1.39	79,647
1994	60,000	54,200	1.35	73,170
1995	61,200	58,275	1.30	75,758
1996	62,700	62,700	1.24	77,748
1997	65,400	63,500	1.17	74,295
1998	68,400	63,500	1.11	70,485
1999	72,600	68,700	1.06	72,822
2000	76,200	74,000	1.00	74,000
2001	80,400	76,500	1.00	76,500

Social Security
Maximum Contribution of OASI
Paid By Employee and Employer that each Pays

Year	Total S.S Tax Rate	OASI Tax Rate		Maximum Wages Taxed		OASI Maximum Annual Contributions (each Pays)
1937 -1949	1.0	1.0	X	$3,000.00	=	30.00
1950	1.5	1.5		3,000.00		45.00
1951-1953	1.5	1.5		3,600.00		54.00
1954	2.0	2.0		3,600.00		72.00
1955-1956	2.0	2.0		4,200.00		84.00
1957-1958	2.25	2.0		4,200.00		84.00
1959	2.5	2.25		4,800.00		108.00
1960-1961	3.0	2.75		4,800.00		132.00
1962	3.125	2.875		4,800.00		138.00
1963-1965	3.625	3.375		4,800.00		162.00
1966	4.2	3.5		6,600.00		231.00
1967	4.4	3.55		6,600.00		234.30
1968	4.4	3.325		7,800.00		259.35
1969	4.8	3.725		7,800.00		290.55
1970	4.8	3.65		7,800.00		284.70
1971	5.2	4.05		7,800.00		315.90
1972	5.2	4.05		9,000.00		364.50
1973	5.85	4.30		10,800.00		464.40
1974	5.85	4.375		13,200.00		577.50
1975	5.85	4.375		14,100.00		616.88
1976	5.85	4.375		15,300.00		669.38
1977	5.85	4.375		16,500.00		721.88
1978	6.05	4.275		17,700.00		756.68
1979	6.13	4.33		22,900.00		991.58
1980	6.13	4.52		25,900.00		1,170.68
1981	6.65	4.70		29,700.00		1,395.90
1982	6.7	4.575		32,400.00		1,482.30
1983	6.7	4.575		35,700.00		1,633.25
1984	7.0	5.2		37,800.00		1,965.60
1985	7.05	5.2		39600.00		2,054.20
1986	7.15	5.2		42,000.00		2,184.00
1987	7.15	5.2		43,800.00		2,277.60
1988	7.51	5.53		45,000.00		2,488.50
1989	7.51	5.53		48,000.00		2,654.40
1990	7.65	5.6		51,300.00		2,872.80
1991	7.65	5.6		53,400.00		2,990.40
1992	7.65	5.6		55,500.00		3,108.00
1993	7.65	5.6		57,600.00		3,225.60
1994	7.65	5.6		60,600.00		3,393.60
1995	7.65	5.6		61,200.00		3,427.20
1996	7.65	5.6		62,700.00		3,511.20
1997	7.65	5.6		65,400.00		3,662.40
1998	7.65	5.6		68,400.00		3,830.40
1999	7.65	5.6		72,600.00		4,065.60
2000	7.65	5.6		76,200.00		4,267.20
2001	7.65	6.2		80,400.00		4,984.80

Is Social Security A Good Deal?

     If a worker started working 45 years ago in 1956 the maximum amount they and their employer have paid into retirement for the employee is $ 141,343.00.
     How much would a worker who has consistently made the Maximum Earnings have, when they retire, had they put the money each year into an investment earning 7 1/2 % compounded yearly?
     Starting with the first year, 1956, the worker had $84.00 deducted from their pay check. Add the same amount their employer paid equals $ 168.00. Using the formula from page 9:19 of this book, would yield $ 4,352.01 if the $168.00 had been invested at 7 1/2% compounded yearly for 45 years.

Calculate 168 X 1.075 Y^x Key 45 = $ 4,352.01

Using the same formula for each of the 45 years, (the worker had an ever increasing amount taken out of their check), the total for the 45 years would be $ 384,204.00

     Now that the employee is retired, how much could they take out each month from a fund had they taken the $ 384,204 and invested it at 7 1/2 % compounded monthly? We will assume the worker is going to live for 30 more years and does not want any money left in the fund after they die. Using the formula of page 9:41 would give the answer of $ 2,686.41 per month for 30 years. Taking the monthly amount times 360 months equals $ 967,107.60 that the employee would receive from the retirement fund if he lives for 30 more years.

     Now let us figure how much Social Security Retirement will pay him for 30 years.

The maximum benefit a person can receive in 2002 from Social Security is $ 1,660.00 per month. To figure the total amount they will receive we will assume Social Security will pay a cost of living increase each year of 3 % per year.

Is Social Security A Good Deal?

Total Pay Out of Social Security for 30 Years
Cost of living increase 3 % per year
$1,660.00 per month X 12 = $19,920.00 for the 1st year

$$\text{1st Year of Retirement} \; X \; \frac{(1 + \text{Cost of Living \%})^{\text{Number of years}} - 1}{\text{Cost of Living \%}}$$

$$\$\,19{,}920 \; X \; \frac{(\,1.03\,)^{30} - 1}{.03}$$

$$19{,}920 \; X \; \frac{2.4272625 - 1}{.03}$$

$$19{,}920 \; X \; \frac{1.4272625}{.03}$$

19,920 X 47.575416 = $947,702.28 would be the total amount a person would receive from Social Security if they were retired for 30 years. The average monthly amount over the 30 year period would be $2,632.50

Comparing the Private Investment to Social Security, per month over a 30 year retirement program.

If you made - Maximum Wages or Above
Private investment at 7 1/2% would have paid $ 2,686.41.
Social Security would have paid $2,632.50 average per year.
A person who makes and contrubutes the maximum amount , would have to invest the same amount from the first year worked and though out their retirment at about 7.5 % to equal the amount paid by Social Security.
If you made - One half of Maximum Wages
Private Investment at 8.7 % would have paid $ 1873.50
Social Security would have paid $1,896.67 average per year.
To equal benefits paid by Social Security you would have to invest at about 8.7 % if you save the same amount that they took out of your pay check.

House

Electric Bills

To determine how the electric company figures your bill you must first determine the amount of kilowatts you used. This is listed on your bill as KWH.

For example we will use the following bill

Meter readings	Prior	Present	Difference (KWH used)	Amount owed
	72506	73536	1030	$92.80

You are not usually charged the same amount for each Kilowatt hour used. Most electric companies charge less for a minimum amount and more at a higher rate. In other words, if you are a big user of electricity in your home, you will pay more for the last KWH used than for the first KWH used. As in the above example, the total of $92.80 was figured as follows.

Lifeline	330 KWHRS	at $0.06485	=	$ 21.40
	210 KWHRS	at $0.08430	=	$ 17.70
	490 KWHRS	at $0.10959	=	$ 53.70
Total	1030			$ 92.80

Note that the break down was 330 KWHrs. at the lowest rate, the next 210 KWHrs at a higher rate and any amount over 540 (330 plus 210) KWHrs was at the highest rate.

Because the last Kilowatt hours are billed at a higher rate, you can save considerably on your bill, by not using as much at the higher expensive rate. For example, if the family in our example were to decrease their usage 15% they would save more than that percent on their bill.

The percent would be taken off the total of 1030

$$1030 \times .15 = 154.5 \text{ KWHrs.}$$

Now figure the dollar amount saved. You base the decrease from the amount of 154.5 from the last KWHrs. used.

$$154.5 \times .10959 = \$16.93 \text{ savings}$$

When compared to the total bill of $92.80

$$\frac{16.93}{92.80} = .182 \text{ or stated as } 18.2 \text{ \% savings}$$

House

British Thermo Unit (Btu)

Btu is the amount of heat to increase 1 pound of water 1 degree Fahrenheit.

> 1000 lbs. of water heated from 55 degrees to 160 degrees
> 1000 X (160 - 55) = Btu
> 1000 X 105 = 105,000 Btus

To determine the Btus to heat 40 gallons of water (water weighs 8.5 lbs. per gallon)
> 40 gallons of water equals (8.5 lbs. X 40) 340 lbs.
> 340 lbs. of water heated from 60 degrees to 180 degrees
> 340 X (180 - 60)
> 340 X 120 = 40,800 Btus

A Leaking Water Faucet.

A gallon of water heated by an electric water heater consumes 0.25KWH.
A hot water facet leaking 120 drops per minute leaks 400 gallons per month
> 400 gallons X 0.25 = 100 KWH
> 100 KWH X .05379 (cost per kilowatt) = $5.39 per month on the electric bill

Cord of Cut Wood for Heating

> A cord of wood is 128 cubic feet, typically measured as 8 feet long, 4 feet high and 4 feet wide.
> A cord foot is 16 cubic feet and is measured as 4 feet long, 4 feet high and 1 foot wide.

House

Wallpaper

Wallpaper comes in a " bolt ." Each bolt contains 2 "rolls" of paper. The bolt is a continuous roll of paper, the length of 2 rolls. When you price wallpaper they typically quote the price for a <u>roll</u> of paper, in other words for half of the bolt. When they size a room, they quote how many rolls will be needed, however they only sell by the bolt.

The typical sizes of a <u>bolt</u> of wallpaper are-
  20 1/2 inches wide X 33 feet long
  21 inches wide by 33 feet long
  27 inches wide X 27 feet long

Measuring a Wall for Wallpaper

Example- The total number of feet of all of the walls of a room is 48 feet and the height is 7 ft. 10 inches

We will first determine how many strands of paper are needed, then account for height then reduce the amount for areas not papered.

Step 1 Change the length in feet into inches.
    48 X 12 = 576

Step 2 Divide the length in inches by the width of the paper used.

    576 divided by 20.5 inches = 28.1
    This means you will need 28.1 strands of wall paper. Change this to 29 strands.

Step 3 To determine how many strands you can get out of a bolt of paper divide the length of the paper by the height of the wall.
    7 ft. 10 inches = 7.83 ft.
    33 feet divided by 7.83 = 4.2 strands   Count as 4 strands. You do not want a seam at the end of the roll (going across the paper) so you will set the extra amount aside, possibly to be used for window sills or whatever.   (*continued*)

House

Measuring a Wall for Wallpaper *(continued)*

Step 4 Divide the number of strands you will get from
a bolt into the total number of strands needed

29 divided by 4 = 7.25 bolts

Step 5 Total the square inches of the areas not to be
covered .
In this example, 2 doors and a window equals
10,800 sq. inches. A bolt covers 8,118 sq
inches so divide into 10,800 equaling 1.33
bolts. 7.25 minus 1.33 = 5.9 . 6 bolts will
barely get the job done and the last thing you
want is to be short, so you should order 7 bolts.

Also be concerned as to the repeat of the pattern
which can require more paper.

Step 6 Please remember that they quote the price of
wallpaper by the roll so you will need 14 rolls.

*Please note that if you measured the square feet of the walls, it
would show that you need less wall paper. The wall is 376
square feet. Each bolt covers 56 square feet. By dividing 376
by 56 = 6.7 bolts . Then if you subtract for the areas not
covered you would only need 5.375 bolts. The last thing you
want is to be short of wallpaper because the next lot will be a
different color.*

Measuring high walls for wallpaper- with no seam in
the middle of the wall (end of roll)

Example- The total number of feet of all the walls of
this room is 56 feet and the walls are 12 feet
high. There are 2 door and 2 windows, not to be
covered, equaling 84 square feet.

Step 1 Change the length in feet into inches.
56 X 12 = 672   *(continued)*

Measuring high walls for wallpaper *(continued)*

Step 2 Divide the total length in inches by the width of the paper used.

   672 divided by 20.5 inches = 32.78
   This means you will need 32.78 strands of wall paper. Change this to 33 strands.

Step 3 To determine how many strands you can get out of a bolt of paper divide the length of the paper by the height of the wall.

   33 feet divided by 12 = 2.75 strands. Count as 2 strands because you do not want a seam part way down part of the wall (at the end of a bolt), so you will set the extra amount aside, possibly to be used for window sills or inside closets.

Step 4 Divide the number of strands you will get from a bolt into the total number of strands needed.

   33 divided by 2 = 16.5 bolts

Step 5 Total the square inches of the areas not to be covered .
   In this example, 2 doors and 2 windows equal 12,096 sq. inches. A bolt covers 8,118 sq. inches so divide into 12,096 equaling 1.49 bolts. 16.5 minus 1.49 = 15. 15 bolts will barely get the job done and the last thing you want is to be short, so you should order 16 bolts. (16 bolts X 2 = 32 rolls)

   Also be concerned as to the repeat of the pattern which can require more paper.

Step 6 Please remember, they quote the price of wallpaper by the roll so you will need at least 32 rolls.

House

Carpet - Measuring

Carpet is sold by the square yard. There are 9 square feet in a yard of carpet.

Example - How many yards are needed for a room 25 feet 9 inches by 13 feet 5 inches?

Change the inches to a decimal.

$\frac{9}{12} = 0.42$

25 ft 9 in. = 25.75 and 13 ft 5 in. = 13.42

Multiply them 25.75 X 13.42 = 345.57 square feet

$\frac{5}{12} = 0.75$

There are 9 square feet in a square yard so divide 9 into 345.57

$$\frac{345.57}{9} = 38.4 \text{ square yards}$$

Measuring a Room for Paint or Wallpaper

To measure the surface area of the walls using the example room, do the following.

Total the perimeter of the room. 25 +25 + 15 + 15 = 80 feet

Multiply times the height of the walls 8 X 80 = 640 square feet

Find the area not to be covered, door 3X7= 21; closet 5 X7=35; 2 windows 2X4X3=24; window 3X3=9. Add them up 21+35+24+9=89.

Subtract from the wall total. 640 - 89 = 551 square feet.

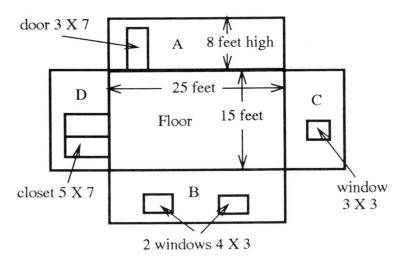

door 3 X 7

A   8 feet high

D

25 feet

C

Floor   15 feet

closet 5 X 7

B

window 3 X 3

2 windows 4 X 3

Paint

Each gallon can of paint usually tells how many square feet it will cover. This is usually 300 to 400 square feet for a gallon of paint. However, if the surface has not been painted for a long time or if it is a rough surface, then it may only cover 150 to 250 square feet. You should also consider how dry the air is when you are going to paint. If the surface has been in a very dry climate, for example winter time and the air has been heated, the relative humidity could be very low causing the surface to soak up more paint. This could use as little as 5 percent to as much as 20 percent more paint.

Example-

Using the previous example room with a wall surface of 551 square feet, divide 400 square feet per gallon into 551 equaling 1.3775 gallons, round of to 1.5 gallons. If the walls have not been painted for a long time, estimate that 50 % more will be needed, then 1.3775 X 1.50 = 2.066 gallons will be needed. If the relative humidity is near 20 % then 15 % more paint may be needed. 2.066 X 1.15 = 2.376 or 2.5 gallons will be required to cover the walls.

House

Measuring a Mound of Topsoil or Sand

Example - You ordered 2 cubic yards of topsoil and it
was delivered and deposited on your driveway.
Is it really 2 cubic yards?

Step 1 Shape the soil into a cone. Measure the
diameter and height.

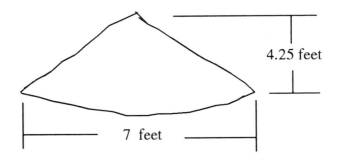

4.25 feet

7 feet

Step 2 Divide the diameter by 2 to find the radius.

7 feet dia. = 3.5 radius

Step 3 Do the following formula (*This formula can also be
found in the Geometric Shapes Chapter*).

1/3 X Pi X R^2 X Height

1/3 X 3.1416 X (3.5 X 3.5) X 4.25

1/3 X 3.1416 X 12.25 X 4.25

1/3 X 3.1416 X 52.06

.333 X 163.55 = 54.5 cubic feet

Change Cubic feet to cubic yards by dividing
54.5 by 27 cubic feet to the yard = 2.02 cubic
feet

Direct Current - D.C.

E = Energy - Voltage
I = Ampere - Current
R = Ohms - Resistance

$$I = \frac{E}{R}$$

$$I = \frac{24}{50} = 0.48 \text{ amps}$$

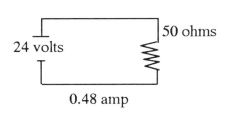

24 volts

50 ohms

0.48 amp

$E = I R$

$E = 0.48 \times 50 = 24 \text{ volts}$

$$R = \frac{E}{I}$$

$$R = \frac{24}{0.48} = 50 \text{ ohms}$$

Batteries in Series Circuit

When batteries are put in series (negative to positive) they total the sum of the voltages of each battery.

$1.5 + 1.5 = 3 \text{ volts}$

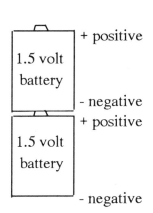

+ positive

1.5 volt battery

- negative
+ positive

1.5 volt battery

- negative

Direct Current (continued)

- When Resistance (motors,
solenoids, resistors, ...)
are in series, in a direct current
circuit, the sum of each items
resistance is the total resistance
of the circuit.
-Current (I) is the same
throughout the circuit.
-The total of the voltage drops
of each resistor equals the
applied voltage.

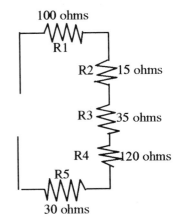

100 ohms
R1
R2 15 ohms
R3 35 ohms
R4 120 ohms
R5
30 ohms

$$I = \frac{E}{R_t} = \frac{120 \text{ volts}}{100 + 15 + 35 + 120 + 30} = \frac{120}{300} = 0.4 \text{ amps.}$$

$$E = I R = 0.4 \times 300 = 120 \text{ volts}$$

$$R = \frac{E}{I} = \frac{120}{0.4} = 300 \text{ ohms}$$

Energy drop over each resistor

$$E = I R$$
$$E_1 = I R_1 = 0.4 \times 100 = 40 \text{ volts}$$
$$E_2 = I R_2 = 0.4 \times 15 = 6 \text{ volts}$$
$$E_3 = I R_3 = 0.4 \times 35 = 14 \text{ volts}$$
$$E_4 = I R_4 = 0.4 \times 120 = 48 \text{ volts}$$
$$E_5 = I R_5 = 0.4 \times 30 = 12 \text{ volts}$$
$$\text{Total} = 120 \text{ volts}$$

Parallel Circuits

When batteries are put in parallel the total voltage output will still be no more than the average of the batteries.

$$\frac{1.5 + 1.5}{2} = 1.5 \text{ volts}$$

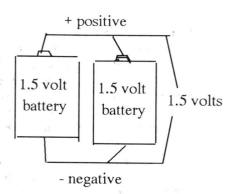

-Every branch of a parallel circuit has the same amount of voltage.
-The total current of the circuit is the total current of each branch when added together.
-Resistance of each branch of a parallel circuit is more than the total resistance of the circuit.

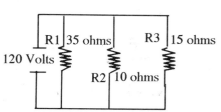

$$R \text{ total} = \frac{1}{\dfrac{1}{R1} + \dfrac{1}{R2} + \dfrac{1}{R3}} = \frac{1}{\dfrac{1}{35} + \dfrac{1}{10} + \dfrac{1}{15}} =$$

$$\frac{1}{0.02857 + 0.10 + 0.06666} = \frac{1}{0.19523} = 5.12216 \text{ ohms}$$

Resistance total = 5.12216 ohms

Electricity

Alternating Current - AC - House Current or Voltage

Typical voltage in a house usually runs from 108 volts to 124 volts. Most houses also have 208 to 230 volts available for dryers, electrical stoves, furnaces and water heaters. However the latter two are usually ran by gas. With all of the available appliances today "popping the circuit breaker ," is a well known occurrence. Older homes "blow a fuse".

Most circuit breakers or fuses are 20 amp 115 volts.
and 15 amp. 115 volts.
Most appliances are rated by power, the amount of watts they use or by the amount of amps (current) they draw.

To determine the amount of amps a 100 watt light bulb draws do the following.

divide watts by volts $\dfrac{watts}{volts}$ $\dfrac{100}{115}$ = 0.87 amps.

To determine the amount of watts, multiply amps. times volts.

$$0.87 \ X \ 115 \ = \ 100 \text{ watts}$$

Typical watt and amp. ratings of appliances around the house.

	Watts	Amps.
Can opener	150	1.3
Toaster	925	7.5
Coffee Maker	1,025	8.5
TV 27 inch	150	1.3
Micro Wave Oven	1,600	13.9
Refrigerator	690	6
Electric Stove	6000	27 (220 Volt)
Washer	1300	11.5
Dryer Electric	4500	20 (220 volt)
Computer & Monitor	175	1.6
Stereo	280	2.6

Electricity

AC Motors

Most AC motors are rated by horse power. If you take the typical formula, Watts = Volts X Amps. you will find that the answer will be much higher than reality. For example, a 1/2 horse power motor at 115 volts draws 8.8 amps. Using the typical formula, 115 X 8.8 = 943 watts. However, 746 watts equals 1 house power. Motors vary as to their efficiency.

A 1/3 horse power, 115 volt capacitor start motor draws 6.6 amps. A high efficiency 1/3 horse power, 115 volt capacitor start, capacitor run motor draws 3.8 amps.

Cost / Efficiency / Return on Investment

Using the above mentioned motor, how long would it take to save enough electricity to make the more efficient motor worth installing?

6.6 amps. minus 3.8 = 2.8 amps.
2.8 amps X 115 volts = 322 watts
A Kilowatt is 1,000 watts, so the above motor saves 0.322 kilowatts per hour.
If a kilowatt cost 8 cents then
0.322 X .08 = 0.02576 dollars saved per hour
If the higher efficiency motor cost $35.00 more then
$35.00 divided by 0.02576 = 1,359 hours.
If the motor is used 6 hours a day then
1,359 divided by 6 = 226.5 days of use and then the more efficient motor has paid for itself.

Spa Usage / Cost

Total Watts used divided by 1,000 X Cost per Kilowatt
Example-   = Cost

A typical Spa has
2.5 H.P. Motor X 746 watts = 1865 watts
1 H.P. Motor   X 746 watts =   746 watts
Heater 5,500 watts   = 5500 watts
Total = 8,111 watts

8.1 kilowatt X $ 0.10 = $ 0.81 per hour used

Electricity

Thickness/Amperage of Wire

Wire Size No.	Ampere Capacity	Size in inches
18	10	0.403
16	15	0.5082 thickness of 1 dime
14	20	0.641 thickness of a penny
12	30	0.0808 thickness of a nickel
10	40	0.1019 thickness of 2 dimes
8	55	0.1285
6	80	0.184

Framing a Picture

The typical frame of 4 sides.

$$\frac{360^0 \text{ in a circle}}{4 \text{ sides}} = 90^0$$

all 3 angles of a triangle must equal 180⁰

$180 - 90 = 90^0$ for two angles

$$\frac{90^0}{2 \text{ other angles}} = 45^0$$

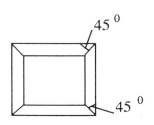

Cutting Angles for a Multi-Sided Shape (Polygon)

This shape could be for a stained glass window or a gazebo. The formula works for any multi-shaped item. You can also find more information and formulas in the Geometric Shapes Chapter of this book.

To find the degree of each angle:

Step 1 Divide the number of sides into 360 ⁰. This gives you the inside angle.

Step 2 Because all triangles have to equal 180⁰ subtract the inside angle from 180.

Step 3 Divide this number by 2 to get the size of the other to angles.

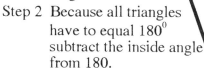

Step 1 $\dfrac{360^0}{5 \text{ sides}} = 72^0$ of one angle

Step 2 $180 - 72 = 108^0$

Step 3 $\dfrac{108}{2} = 54^0$

Construction

Outer Length of a Side of a Polygon

$$\frac{\left(\begin{array}{c}\text{Number}\\\text{of Sides}\end{array} \text{X} \quad \text{Diameter}\right) \text{X sine of} \left(\dfrac{360}{\begin{array}{c}\text{number}\\\text{of sides}\end{array}} \text{X} \dfrac{1}{2}\right)}{\text{All of the above divided by the number of sides}}$$

Example - What is the length of the outside edge of the example polygon?

Outside Length ?

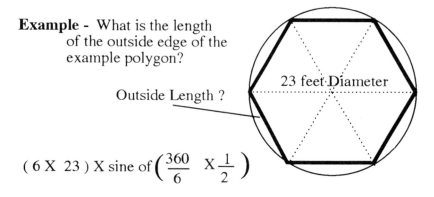

23 feet Diameter

$$(6 \text{ X } 23) \text{ X sine of} \left(\frac{360}{6} \text{ X} \frac{1}{2}\right)$$

$$(6 \text{ X } 23 \text{ X sine of} \left(60 \text{ X} \frac{1}{2}\right) = 6 \text{ X } 23 \text{ X sine of } 30$$

on a calculator enter 30 then the sine key and 0.5 will appear

138 X 0.5 = 69   divide this number by the number of sides

$$\frac{69}{6 \text{ sides}} = 11.5 \text{ feet}$$

The outside length is 11.5 feet

Stress and Strength

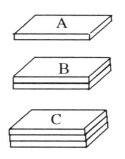

In this illustration each set of boards, if they are not bonded together, have the strength as to the number of boards used.
(C is 3 times stronger than A)
If they are bonded (glued) together they are much stronger. If boards B are bonded together they are 4 times stronger than A. The 3 boards of C bonded together have the strength of 9 boards.

The strength of a board is proportional to the square (multiplied times itself) of the depth of the board.

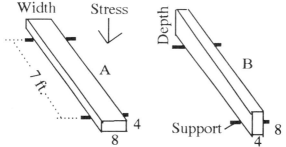

**Beam
Load Strength**

$$\text{Load Strength} = \frac{\frac{\text{Stress rating of}}{\text{wood used}} \text{ X Width X Depth X Depth}}{9 \quad \text{X Length} \quad \text{in feet}}$$

(A standard No.)

Example - What is the load strength in the middle of Board B with wood that has a Stress Rating of 1,600f ?

$$\text{Load Strength} = \frac{1,600 \text{ X } 4 \text{ X } 8 \text{ X } 8}{9 \text{ X } 7}$$

"f" is a grading system used by the building trade.

$$\text{Load Strength} = \frac{409,600}{63} = 6,502 \text{ pounds can be supported}$$

as long as the weight is distributed evenly

If the weight is going to be in the middle divide 6,502 in half = 3,251 lb.

Construction

Beam Load Strength (continued)

Example- The same board set on its side (example A)

$$\text{Load Strength} = \frac{1,600 \times 8 \times 4 \times 4}{(\textit{Standard No.})\ 9 \times 7}$$

Load Strength = $\frac{204,800}{63}$ = 3,251 pounds can be supported as long as the weight is distributed evenly

Example - If we glue a board to the top of Example A, that is the same size, we have the following.

$$\text{Load Strength} = \frac{1,600 \times 8 \times 8 \times 8}{9 \times 7}$$

Load Strength = $\frac{819,200}{63}$ = 13,003 pounds can be supported as long as the weight is distributed evenly

This combination is 4 times stronger with just one board added.

Beam Load Strength

Example - You have a deck that meets code for supporting the weight it is presently used for. You now want to add a Hot Tub weighting 4,000 pounds including 3 people and water in it. The Hot Tub measures 7.5 feet X 7.5 feet.

View looking down on the deck

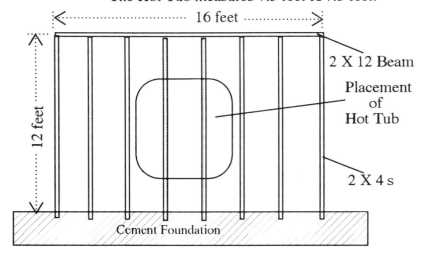

How many 2 X 8 s with an f factor of 1600 will you have to add to the present structure to properly support the additional weight of the Hot Tub ?
Use the actual size of the lumber (1.625 by 7.5).

$$\frac{1600 \times 1.625 \times 7.5 \times 7.5}{9 \times 12 \text{ feet}} = \frac{146,250}{108} = 1,354 \text{ pounds}$$

a standard no.

Now divide 1,354 by 2 because the Hot Tub will be in the middle equaling 677 pounds.

$$\frac{4,000}{677} = 5.9 \text{ or } 6$$

You will need 6 each 2 X 8 s added to the present structure to support the Hot Tub.

Construction

Beam Load Strength (continued)

> **Example** - Using the previous drawing, how much
> stronger would the 2 X 12 have to be, to
> support the Hot Tub?
>
> We will first try another 2 X 12
>
> $$\frac{1600 \text{ X } 1.625 \text{ X } 11.5 \text{ X } 11.5}{9 \text{ X } 16} = \frac{343,850}{144} = 2,387$$
>
> Now divide into the weight of the Hot Tub

weight of Tub $\frac{4,000}{2,388}$ = 1.67 boards of 2 X 12 needed to add to
the present 2 X 12 beam.
(2 boards will be needed)

> We will now try a 4 X 12
>
> $$\frac{16,000 \text{ X } 3.625 \text{ X } 11.5 \text{ X } 11.5}{9 \text{ X } 16} = \frac{767,050}{144} = 5,327$$
>
> 5,326 pounds is more than what the Hot Tub weights,
> so one 4 X 12 is more that adequate.

Please note that when you first look at the comparison of a
2 X 12 and a 4 X 12 you would think the 4 X 12 would only be
twice as strong as one 2 X 12. However the finish size of a 4 X
12 is 3.625 inches thick and double the thickness of a 2 X 12 is
3.25 inches.

Column Load Strength

When determining the amount of weight a column can take,
you look at two factors, the weight it can take before it is
crushed and the weight it can take before it bends and breaks.
the following is as accepted standard of both conditions.

Column Load Strength (continued)

Example - How many pounds can a
board support that is
4 X 4 inches and
6 feet long ?
The lumber company rates the
board as having **1,500,000**
Modules of Elasticity.

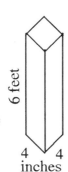

6 feet

4   4
inches

Allowable load
pounds per =
square inch

*0.3 is a standard
number*

$$\frac{0.3 \text{ X Elasticity Rating}}{\left[\left(\dfrac{\text{Length X Length}}{\text{Width X Width}}\right)\right]^2}$$

Allowable load
pounds per =
square inch

$$\frac{0.3 \text{ X } 1{,}500{,}000}{\left[\left(\dfrac{6 \text{ X } 6}{4 \text{ X } 4}\right)\right]^2}$$

Allowable load
pounds per =
square inch

$$\frac{450{,}000}{\left[\left(\dfrac{36}{16}\right)\right]^2}$$

Allowable load
pounds per =
square inch

$$\frac{450{,}000}{\left[\dfrac{36}{8}\right]^2}$$ *36 divided by 8 = 4.5*

Allowable load
pounds per =
square inch

$$\frac{450{,}000}{4.5 \text{ X } 4.5} = \frac{450{,}000}{20.25} = 22{,}222 \text{ pounds}$$

This board can hold 22,222 pounds before it is crushed or bends.

Construction

Board Feet

A Board Foot is 1 foot wide 1 foot long and 1 inch thick. The nominal size is used. If a board is less than 1 inch thick it is considered as being 1 inch. Example - A 1 inch thick X 6 inch wide X 20 feet long board. Change the 6 inches to 0.5 because it is one half of a foot.   1 X .5 X 20 = 10 board feet

Size of Lumber

Nominal Size	American Standard Finished Size
1 X 4	25/32 X 3 5/8
1 X 6	25/32 X 5 5/8
1 X 8	25/32 X 7 1/2
1 X 12	25/32 X 11 1/2
2 X 2	1 5/8 X 1 5/8
2 X 4	1 5/8 X 3 5/8
2 X 6	1 5/8 X 5 5/8
2 X 8	1 5/8 X 7 1/2
2 X 10	1 5/8 X 9 1/2
2 X 12	1 5/8 X 11 1/2
4 X 4	3 5/8 X 3 5/8
4 X 6	3 5/8 X 5 5/8
4 X 8	3 5/8 X 7 1/2
4 X 12	3 5/8 X 11 1/2
6 X 6	5 5/8 X 5 5/8
6 X 8	5 5/8 X 7 1/2
6 X 12	5 5/8 X 11 1/2
6 X 18	5 5/8 X 17 1/2
8 X 8	7 1/2 X 7 1/2
8 X 16	7 1/2 X 15 1/2
10 X 10	9 1/2 X 9 1/2
10 X 16	9 1/2 X 15 1/2
12 X 12	11 1/2 X 11 1/2
14 X 14	13 1/2 X 13 1/2

Bricks

To determine the number of bricks needed to build a wall, measure the area of the brick that will be facing outward.

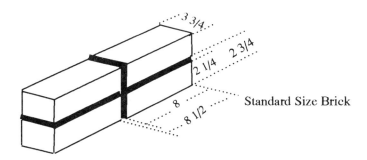

Standard Size Brick

Find the area represented by 2 1/4 X 8, then add 1/2 inch for the mortar to each of the dimensions, = 2 3/4 X 8 1/2. Change the sizes to decimals, 2.75 X 8.5 = 23.375 square inches.

Now find the amount of square feet for the wall you are going to build. Our example is 4 feet high and 25 feet long. 4 X 25 = 100 square feet.
A square foot has 144 square inches (12X12 = 144).
Multiply 144 X 100 square feet = 14,400 square inches for the wall.
Divide the number of square inches of the brick with mortar into 14,400.

$$\frac{14,400}{23.375} = 616 \text{ Bricks}$$

The design will probably be staggered and extra bricks will be needed. Determine how many rows will be needed.
4 feet high X 12 inches = 48 inches $\frac{48}{2.75}$ = 17.45 rows

Change 17.45 to 18 and divide by 2 = 9 extra bricks
(there will be 18 half pieces)

Construction

Cubic Yards

A Cubic Yard is 3 feet high by 3 feet wide and 3 feet deep. It is 27 cubic feet (3 X 3 X 3 = 27 cubic feet) or (36 X 36 X 36 = 46,656 cubic inches).

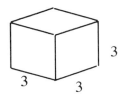

Example - How many Cubic Yards of cement are needed for a patio 30 feet by 20 feet and the cement will be 4 inches thick? Only one side will have a footing 6 inches wide and 8 inches deep.

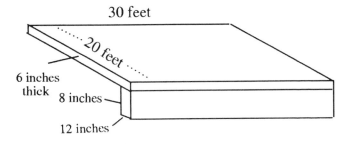

30 feet

20 feet

6 inches thick

8 inches

12 inches

Change 6 inches to a percent of a foot = 0.5

Main slab -   30 X 20 X 0.5 = 300 cubic feet
Divide by 27 to get cubic yards = 11.1 cubic yards

Footing -   8 inches X 12 inches X (30 feet X12 inches)
96 X 360 = 34,560 cubic inches
To change cubic inches to cubic yards, divide cubic inches by 46,656.
34,560/ 46,656 = 0.74 cubic yards

11.1 + 0.74 = 11.84 cubic yards of concrete are needed.

Cubic Yards (continued)

You can not make perfectly square corners when digging a foundation. Therefore, it would be wise to add 10 % to the amount for the footing and 5 % more for the slab.

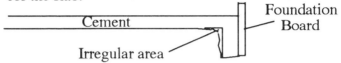

7.4 X .05 = .37 cubic yard
.37 X .10 = .037 cubic yard
Add then all together   7.4
                          .37
                         ..37
                         .0.037
                        8.177 cubic yards

Mixing Concrete

1 part cement
2 parts sand
3 parts gravel
1 / 2 part water

Using a 2 cubic foot cement mixer (holds 15 gallons) add the following.

2 gallons of cement
4 gallons of sand
6 gallons of gravel
1 gallon of water

Mixing the above will produce about 11-12 gallons of mixed concrete (1.45 - 1.6 cubic feet)

Construction

Length of Rafters

-Measure the length of the span.
-Determine the desired pitch of the roof. Find the
corresponding Multiplier on Page 14:16.
-Multiply the span by the Multiplier, then divide by 2,
-Add to the length for the overhang and then subtract
one-half the width of the Ridge board.

Example -

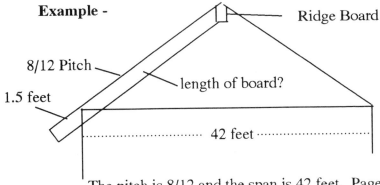

Ridge Board

8/12 Pitch

length of board?

1.5 feet

42 feet

The pitch is 8/12 and the span is 42 feet. Page
14:16 shows the multiplier as 1.2019.

42 X 1.2019 = 50.48 feet

50.48 divided by 2 = 25.24 feet

Add 1.5 feet for the overhang = 26.74 feet.

Subtract 1/2 of the width of the Ridge board

1.625/2 = 0.8125   Rounded off to 1 inch.

1 inch = 0.0833 of a foot

26.74 – 0.0833 = 26.66 feet

To change to feet / inches
0.66 X 12 inches = 7.92 inches round off to 8

24 feet 8 inches

Angle of Cut for Rafters

We will use the example from the previous page. The pitch
was 8/12. Using a large framing square, place it on the rafter
as shown, aligning the 8 and 12 of the framing square to the
board. Make a line from 8 inches to the top of the board. This
is the cut line for the top of the rafter. Now measure from the
bottom of the rafter, where you had the square at 8 inches, to
the other end and mark it at 24 feet 8 inches. Now place the
framing square up side down on the rafter and again line up
the mark with the 8 and the 12 of the framing square (the
reverse of the other end). Mark the cut line.

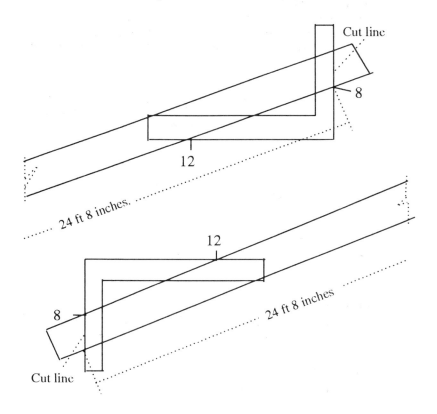

Construction

Pitch of a Roof

Example- 9/12 means that for every 12 inches the roof raises 9 inches.

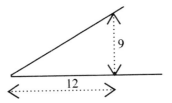

Measuring Pitch

Place a framing square on the edge of the roof as shown. Line up the 12 while keeping the square level. The number you read on the lower part is the raise of the roof.

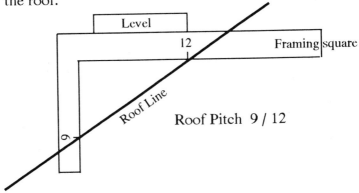

Area of a Roof

You could go up on the roof and actually make all of the measurements; figure the slant (hypotenuse); or use a multiplier.

Using the hypotenuse will first be done, then we will use the multiplier method.

Area of a Roof

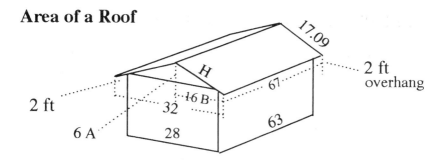

Measure the height of the roof (A)
Measure the width of the house.
Measure the length of the house.
Add the eaves to the above measurements.
Example - What is the square footage of the roof on a
house that measures 63 X 28 ?

Height of the roof is 6 feet

Width of house plus eaves
    28   + 2 + 2 = 32 divided in half = 16
Length of house plus eaves
    63   + 2 +2 = 67

Do the following formula for finding the
hypotenuse then the area.

$$H = \sqrt{A^2 + B^2} = \sqrt{6^2 + 16^2} = \sqrt{(6 \times 6) + (16 \times 16)}$$

$$H = \sqrt{36 + 256} = \sqrt{292} \quad \text{calculate } 292 \; \boxed{\sqrt{\text{ key}}} = 17.09$$

17.09 X 67 = 1,145.03 sq. ft. X 2 for the other side = 2,290.06
Square feet

Construction

Multiplier Method for Area of a Roof

Use the previously drawn house.

Step 1 We will use the height and half the width. 6/16
However pitches are stated as *variable* / 12
(*12 is always standard*)
6/16 = ??/12
12 divided by 16 = .75
Multiply 6 by .75 = 4.5
The pitch of the roof is 4.5 / 12

$$\frac{6}{16} = \frac{???}{12}$$

Step 2 Measure the house length and width, then add
for the overhang of the roof. 63+2+2 = 67
28+2+2 = 32

67 X 32 = 2,144

Step 3 To make up for the slant of the roof, use the
following chart, find 4/12 and 5/12 then add
their 2 corresponding numbers 1.0541 and
1.0833 = 2.1374. Divide 2.1374 by 2 = 1.0687
to get their average or actually 4.5.

Step4 Multiply 1.0687 X 2,144 = 2,291.3 square foot.
roof surface

Roof Pitch Multipliers

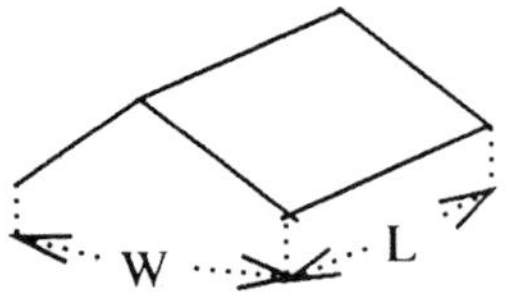

W X L X M= Sq. Ft.

Pitch	Multipliers
16/12	1.6667
15/12	1.6001
14/12	1.5366
13/12	1.4743
12/12	1.4142
11/12	1.3566
10/12	1.3017
9/12	1.25
8/12	1.2019
7/12	1.1577
6/12	1.1181
5/12	1.0833
4/12	1.0541
3/12	1.0308
2/12	1.0138

Pitch without Framing Square

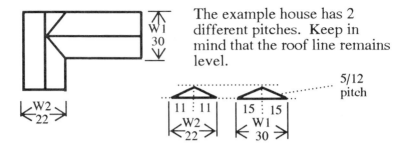

The example house has 2 different pitches. Keep in mind that the roof line remains level.

5/12 pitch

We know from the above problem, the pitch of the wing that is 30 feet wide is 5/12. To find the pitch of the other wing we need to know how high the roof is.

$\frac{5}{12} = \frac{?}{15}$ 15 divided by 12 = 1.25   Multiply 1.25 X 5 = 6.25
The roof height of both wings is 6.25

$\frac{6.25}{11} = \frac{?}{12}$ 6.25 divided by 11 = 0.56818
0.56818 X 12 = 6.81818/ 12 roof pitch

Determining Pitch Multiplier from a Pitch with Decimal Point

Example- Pitch of 6.818 /12

Step 1   Using the previous Chart of Roof Pitch Multipliers, find the pitches for 7/12 and 6/12
7/12 = 1.1577   6/12 = 1.1181
1.1577 minus 1.1181 = .0396

Step 2   From the pitch in question (6.818) delete the 6

Step 2   Multiply .0396 X .818 = .032393

Step 3   Add .032393 + 1.1181 = 1.150496
Rounded off to 1.1505

The Roof Pitch Multiplier for 6.818/12 is 1.1505

Construction

Determining the Multipliers for Roof Pitch

Step 1 Find the slant width by using the formula for finding the hypotenuse of a triangle

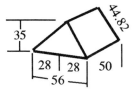

$$\sqrt{(35 \times 35) + (28 \times 28)}$$

$$\sqrt{1225 + 784}$$

$$\sqrt{2009} \quad = \quad 44.82 \text{ feet}$$

Step 2 Divide the width of the roof by 2
56/2 = 28

Step 3 Divide the slant width by one half of the width of the roof.

$$\frac{44.82}{28} = 1.6007$$

Check-

56 X 50 = 2800 2800 X 1.6001 = 4,481.96 square feet
Or
44.82 X 50 = 2241 X 2 for the other side = 4,482
square feet

Making Irregular Roof Pitch Number to Standard Ratio

In the above example the roof pitch is 35/28
However the standard roof pitch is related to how many inches rise every 12 inches. (variable / 12)
35/28 = ??/12
28 divided by 12 = 2.333
35 divided by 2.333 = 15
The roof pitch of the above example is 15/12
Another way
12 divided by 28 = 0.428571
35 X 0.428571 = 14.999 = 15

Determining Height of a Roof

Example- The pitch of the roof is 20/12 and the width of the roof is 56 feet. How high is the roof?

Step 1 Divide the width by 2   56/2 = 28
Step 2 Divide this number by 12

$$\frac{28}{12} = 2.3333$$

Step 3 Multiply that number by the pitch riser number.
2.3333 X 20 = 46.6667

\ *riser number*

The roof is 46.67 feet high.

Area with Joining Roof Lines

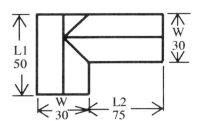

If both ends of the L shaped house have the same roof pitch , there is not a difference as to the area that unites them.

L1 + L2 X W X Multi factor = Square feet

Example- Using the above house with a pitch of 5/12

50 X 30 = 1500   30 X 75 = 2,250

1500 + 2250 = 3,750   *The multiplier for 5/12 pitch is 1.0833*
                        (*see pervious chart*)
3,750 X 1.0833 = 4,062.375 Square Feet

Area of Roof with Joining Roof Lines & Two
Different Pitches

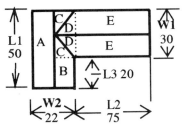

We have already figured
the Multipliers for the
2 different roof lines
W1 = 1.0833 and
W2 = 1.1505

Area A $\dfrac{L1 \times W2}{2}$ X M $\dfrac{50 \times 22}{2}$ X 1.1505 = 632.72

Area B $\dfrac{L3 \times W2}{2}$ X M $\dfrac{20 \times 22}{2}$ X 1.1505 = 253.09

Area C $\dfrac{W1 \times W2}{4}$ X M $\dfrac{30 \times 22}{4}$ X 1.1505 = 189.82

Area D $\dfrac{\frac{W2}{2} \times W1}{2}$ X M $\dfrac{\frac{22}{2} \times 30}{2}$ X 1.0833 = 178.74

Area E   L2 X W1 X M   75 X 30 X 1.08333 = 2437.43

Total square feet   3,691.8

Another Example with a Different Roof Line

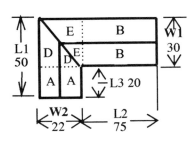

A = 22 X 20 X 1.1505 = 506

B = 75 X 30 X 1.0833 = 2,437

D = $\dfrac{22 \times 30 \times 1.1505}{2}$ = 380

E = $\dfrac{22 \times 30 \times 1.0833}{2}$ = 715

Total Square Feet   4,038

Stairs

Adding the Rise and Run should not equal more than 19 inches or be less than 17 inches. The normal Rise is 7 inches and the Run is adjusted to meet the 19 - 17 rule.

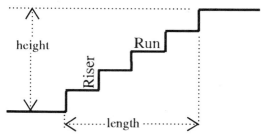

Example - What number of steps are needed for a flight of steps 12 feet high and the of length 18 feet. What should the runs and risers measure?

12 feet X 12 inches = 144 inches $\dfrac{144}{7}$ = 20.57 risers

All of the Risers must be the same height. To get an even number of risers, divide 144 by the approximate number of risers needed (20).

$\dfrac{144}{20}$ =7.2 inches for each Rise and 19 runs are needed.

To determine the size of the runs divide the length by the number of required runs.

18 feet X 12 inches = 216 $\dfrac{216}{19}$ = 11.36 inches

In the above example there will be risers 7.2 inches high, and runs will be 11.36 deep.

Construction
Weight of Building Materials

Boards

2 inch	X 4 inch	X 8 feet	Douglas Fir	Green	17 - 18 pounds
2	X 6	X 8	Douglas Fir	Dry	7 -9 pounds
4	X 6	X 8	Douglas Fir	Green	38 - 42 pounds
2	X 6	X 10	Redwood Dry		15 -18 pounds
2	X 6	X 8	Redwood Dry		10 -12 pounds
1	X 8	X 6	Redwood Dry		5 - 7 pounds

Particle Board

3/4 inch X 4 feet X 8 feet is 90 pounds
Figure 45 pounds per cubic foot

Example 1/2 X 4X 8 = 0.5 X 48 X 96 = 2,304 cubic inches
There are 1728 cubic inches in a cubic foot
2,304 / 1728 cubic inches = 1.333 cubic feet
1.333 X 45 pounds per cubic feet = 60 pounds

Plywood

3/4 X 4 X 8	CDX	66 pounds (lb.)
3/4 X 4 X 8	BCX	72 pounds
1/2 X 4 X 8	CDX	40 pounds
1/2 X 4 X 8	ACX	48 pounds
1/2 X 4 X 8	BCX	54 pounds
5/8 X 4 X 8	Sheathing	66 pounds

Bricks 3 1/2 X 2 X 8 1/2 Clay or Concrete 6 -7 lb.

8 X 8 X 16	Concrete			34 lb.
8 X 16 X 1 5/8	Concrete Stepping Stone			16 lb.
12 X 12 X 2	"	"	"	26 lb.
16 X 16 X 1 1/4	Cobblecrete "		"	26 lb.
12 X 1 3/4 Round	Aggregate Stepping Stone			16 lb.
16 X 1 3/4 Round	"	"	"	27 lb.
16 X 16 X 2	Decorative	"	"	37 lb.

Paint   10 - 12 pounds per gallon

Height of a Flagpole

Sunny day-

Step 1 Take a yard stick or any long object you know the length of and stand it straight up. Measure from its base to the end of its shadow cast by the sun.

Step 2 Measure the length of the shadow from the flagpole.

Step 3 Divide the length of the shadow from the yard stick, into the shadow of the flag pole. Take that number and multiply it by the length of the yard stick.

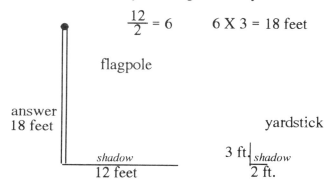

$$\frac{12}{2} = 6 \qquad 6 \times 3 = 18 \text{ feet}$$

flagpole

answer
18 feet

yardstick

3 ft. *shadow*

shadow
12 feet

2 ft.

Cloudy Day or No space to Cast a Shadow

Step 1 Make a right triangle with the base and height having the same length.

same

90⁰

same

Step 2 Put the triangle on a table or stand that can be easily moved. The triangle has to be level so you may need a carpenters level.

Outdoors

Height of a Flagpole (continued)

Step 3 Make a mark on the flag pole, the same
height as the table.

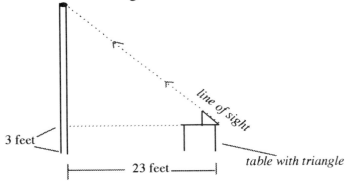

3 feet

23 feet

line of sight

table with triangle

Step 4 Sight along the triangle, moving away or
closer to the flagpole until you align the
triangle with the top of the flagpole.
You must have the triangle level or you
must sight from the triangle to the mark
on the flagpole as well as the top of the
flagpole.

Step 5 After you have done so, measure from
the triangle to the base of the flagpole.
Then add the height of the stand or
table to this distance.

In the above illustration the right edge of the
triangle in 23 feet from the flagpole. The table
is 3 feet high. 23 + 3 = 26 feet. This distance
and the height of the flagpole is the same.

Measuring long distances - Body of Water or Canyon

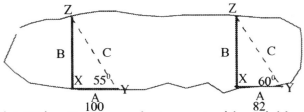

Method A Using Surveyors instrument with variable degrees

From point X sight across the pond to an object designated as
      point Z.
From point X go 100 feet to point Y at a 90 degree angle from
      line B.
From point Y sight to point Z and measure the number of
      degrees from line A to line C. In the example it is 55
      degrees.
To find the length of line B
            B = Line A times the tangent of 55 degrees
            B = 100 X tan of 55
                  calculate 55 │*tan key*│ *= 1.42815*
            B = 100 X 1.42815 = 142.8 feet

Method B Using a standard right triangle with a 60 degree
                                          angle.
From point X sight across the pond to an object designated as
      point Z.
Using the right triangle, sight from point X to point Z and to a
      point beyond Y.
Using a right triangle with a 60 degree angle, walk towards
      point Y sighting down the triangle to X and eventually
      to point Z. When you can sight down the triangle to
      point X and to point Z stop and measure from that point
      to point X. In the example it is 82 feet.
To find the length of line B
            B = Line A times the tangent of 60 degrees
            B = 82 X tan of 60
                  calculate 60 │ *tan key*│ *= 1.7321*
            B = 82 X 1.7321 = 142.03

Outdoors

Acres

An acre is 43,560 square feet or 4,840 square yards. A football field is slightly more than 1 acre.

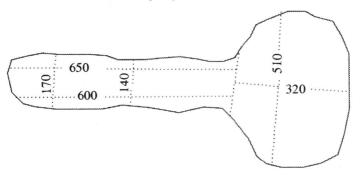

To figure out how many acres the above pond covers

Step 1   Segregate it into 2 different shapes, a rectangle and an oval.

Step 2   The formula for finding the area of an oval is-

Area = Pi X Average Radius^2

Divide the diameters 510 and 320 in-half to get radius

$$Area = 3.1416 \ X \left(\frac{255 + 160}{2} \right)^2$$

Area = 3.1416 X 207.5 X 207.5 = 135,265 Square feet

Step 3   The long part of the pond looks most like a rectangle
The formula for a rectangle is-

Area = length X width
Average the 2 lengths and 2 widths
650 + 600 = 1250 divided by 2 = 625
170 + 140 =   310 divided by 2 = 155

Area = 625 X 155 = 96,875 square feet

Step 4   Add the oval part of the pond to the rectangular part.
135,265 + 96,875 = 232,140 square feet

Step 5   Divide 232,140 by 43,560 square feet = 5.329 Acres

Acre Feet of Water

An acre foot of water, is 1 acre of water 1 foot deep. Using the previous example pond we will make numerous depth measurements to find its average depth. For our example we will say it is 22 feet deep.

5.329 acres X 22 = 117.24 acre feet

Measuring a swimming pool for gallons of water

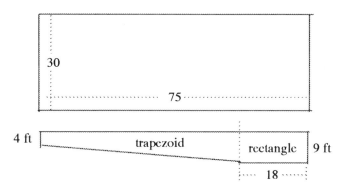

Step 1 Divide the pool into two different sizes, a trapezoid and a rectangle.

Step 2 Determine the volume of the trapezoid section first. The width is 30 feet and the length is (75 - 18) 57 feet The average depth is 4 feet + 9 feet divided by 2 = 6.5 feet

30 X 57 X 6.5 = 11,115 cubic feet

Step 3 Determine the volume of the rectangular section.

18 X 30 X 9 = 4,860 cubic feet

Step 4 Add both sections together

11,115 + 4,860 = 15,975 cubic feet

1 cubic foot holds 7.48 gallons

Step 5 Multiply 15,975 by 7.48 = 119,493 gallons to fill the pool.

Outdoors

Measuring a Kidney Shaped Pool

Do the following steps to
determine how many
gallons it would take to
fill the example pool.

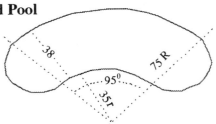

Step 1 Determine the area of the main section (not counting at
this time the two half circle ends).

$$A = \left(\frac{\text{the number of degrees}}{360^0} \times Pi \right) \times (R^2 - r^2)$$

$$A = \left(\frac{95}{360} \times 3.1416 \right) \times \left[(75 \times 75) - (35 \times 35) \right]$$

$$A = (0.264 \times 3.1416) \times (5625 - 1225)$$

$$A = 0.8294 \times 4400 = 3,649.36 \text{ Square feet}$$

Step 2 Determine the area of the 2 ends that are half circles.
We will combine the 2 ends and do the area for a circle.

$$A = Pi \times R^2 \quad \text{To get the radius - one half of 38} = 19$$

$$A = 3.1416 \times 19 \times 19 = 1134.1 \text{ square feet for}$$
both ends

Step 3 Add both of the sections together

Total Area = 3,649.36 + 1134.1 = 4783.46 square feet

Step 4 If the depth of the pool slants down from the shallow
end to the deep end, average the depth. In the example
pool the average depth is 4.5 feet.

4.5 X 4,783.46 = 21,525.6 cubic feet

Step 5 To determine gallons (a cubic ft. contains 7.48 gallons)

21,525.6 multiplied by 7.48 = 161,011.5 gallons

Measuring Fluid Level in a Cylindrical Tank

Please note that when you measure the fluid level, the bottom 2 inches contain less fluid than 2 inches in the middle of the tank.

36 inch diameter 200 gallons

Step 1 Take a measuring rod and measure the level of fluid in the tank. Then determine the percent of the rod that the fluid level covers.   In the example tank the fluid level is 12 inches deep and the tank is 36 inches in diameter. 12/36 = 0.333.

Step 2 Using the chart find a percentage closest to 33%, which is 35%. Go across to .312582.

Step 3 The tank capacity is 200 gallons.
Multiply 200 X 0.312582 = 62.5164
approximate gallons

Percent of Measuring Rod	Percent of Capacity of Tank
.05	.0167685
.10	.0502
.15	.096083
.20	.142917
.25	.193917
.30	.2534
.35	.312582
.40	.376083
.45	.4395815
.50	.50
.55	.5604185
.60	.623917
.65	.687418
.70	.7469
.75	.806083
.80	.857083
.85	.903917
.90	.9499
.95	.983232
1.00	1.00

Outdoors

Water Hose Garden Sprayer

Most Chemical sprayers have a label on them to tell the ratio of mixture to gallons sprayed. However, if the label is off or you question the true ratio it actually sprays, do the following.

1. Fill up the sprayer to the 32 ounce (oz.) mark with plain water. Put the sprayer on the water hose.
2. Turn the water on with the same amount of water coming out that you will actually be using to spray with.
3. Using an empty 5 gallon container, close off the vent hole or turn the lever on, as if you were actually spraying with chemicals and fill up the 5 gallon container.
4. Determine the amount of water that has been used from the chemical bottle of the sprayer during the time it took to fill the 5 gallon container.
    Example: If the level of water is at the 20 oz. level. 32 subtract 20 = 12 oz. were used. The ratio would then be 12 oz. to 5 gallons of water. By dividing 12 by 5 we get 2.4 oz. per gallon. This can be rounded of to 2 1/2 oz. per gallon.
5. If the chemical you are spraying requires 2 1/2 oz. per gallon and the area requires 8 gallons of spray simply pour in 20 oz. of chemical. (2 1/2 X 8 = 20)

If the chemical requires a weaker amount (less than the ratio from your sprayer) do the following.

Example: If the chemical requires the ratio of 1 1/2 oz. per gallon and your sprayer does 2 1/2 oz. per gallon, do the following. Start with the ratio of your sprayer, 2 1/2 and subtract the required ratio of 1 1/2 ratio = 1 oz. Add 1 oz. of water per 1 1/2 oz of chemical. If the area you are to spray requires 5 gallons, multiply the amount of water and chemical by 5 and pour into chemical bottle of the sprayer.
1 oz. of water X 5 = 5 oz.
1 1/2 oz. of chemical X 5 = 7.5 oz.

If the chemical requires a stronger ratio than your sprayer can put out, enlarge the siphon system and re-test.

Belt Length

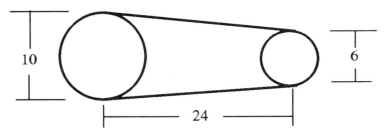

π X Diameter of first pulley divided by 2 =
π X Diameter of second pulley divided by 2 =
Distance from the center of each pulley X 2 =
                                                    Total ──────

3.1416 X 10 = 31.416 divided by 2 = 15.71
3.1416 X 6 = 18.85   divided by 2 =   9.42
                                24 X 2 = 48.00
                                          ─────
                                          73.13
                    Belt Length = 73 inches

Pulley Ratios

Using the above pulley example, if the pulley on the left, the 10 inch pulley, was the drive pulley and the 6 inch pulley was the driven pulley, the ratio would be 0.6 to 1. Stated the normal way 0.6 : 1
                    6 divided by 10 = 0.6
When the 10 inch pulley makes 0.6 of a full turn the smaller 6 inch pulley makes one full revolution.

If the 6 inch pulley is the driver the ratio would be 1.67 : 1   10 divided by 6 = 1.67
When the small pulley is the driver, it takes 1.67 revolutions to turn the larger 10 inch pulley 1 revolution.

Mechanical

Pulley RPM

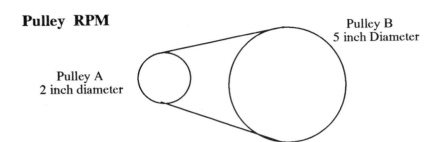

Pulley B
5 inch Diameter

Pulley A
2 inch diameter

$$\frac{\text{RPM of Drive Pulley}}{\left(\dfrac{\text{Diameter of Driven pulley}}{\text{Diameter of Driver pulley}}\right)} = \text{RPM of Driven Pulley}$$

Example- What will be the **RPM** of Pulley B if Pulley A is turning at 520 **RPM**?

$$\frac{520}{\left(\dfrac{5}{2}\right)} = \text{RPM of Pulley B}$$

$$\frac{520}{2.5} = 208 \text{ RPM for Pulley B}$$

Example- If Pulley B is the driven pulley and it is 10 inches in diameter and has to turn at 416 **RPM**, how fast does gear A have to turn if it is 4 inches in diameter?
Treat Pulley B as the driver gear and use the above formula.

$$\frac{416}{\left(\dfrac{4}{10}\right)} = \text{RPM of Pulley A}$$

$$\frac{416}{0.4} = 1,040 \text{ RPM for Pulley A}$$

Pulley Size

Example- You have a motor that turns 1725 revolutions per minute (**RPM**). You have a need to drive a mechanism, 500 revolutions per minute. What size pulley do you need on the mechanism, if the motor has a 3 inch pulley ?

$$\frac{\text{Driver RPM}}{\text{Driven RPM}} \; X \; \frac{\text{Driver Pulley}}{\text{Diameter}} = \text{Driven Pulley}$$

$$\frac{1725}{500} \; X \; 3 \text{ inches} = \frac{\text{Driven Pulley}}{\text{Diameter}}$$

$$3.45 \; X \; 3 \; = \; 10.35 \;\; \text{Diameter}$$

3 inch
Drive pulley

1725
RPM

10.35 inches changed to
10 3/8 inch
Driven Pulley

500
RPM

To check if the answer is correct, multiply the diameter and the **RPM** of both pulleys. The answer should be the same for both.

$$3 \; X \; 1725 = 5175 \quad\quad 10.35 \; X \; 500 \; = 5175$$

Example- You have a machine that has to turn at 500 **RPM**. It has a 9.5 inch pulley and a drive motor that turns at 1725 **RPM**. What size pulley do you need on the drive motor?

$$\frac{\text{Driven pulley size}}{\left(\dfrac{\text{Driver RPM}}{\text{Driven RPM}}\right)} = \text{Driver Pulley Dia.}$$

$$\frac{9.5}{\left(\dfrac{1725}{500}\right)} \;=\; \frac{9.5}{3.45} \;=\; 2.75 \text{ Dia. of Driver Pulley}$$

Mechanical

Compound Drive System

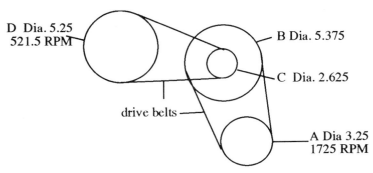

D Dia. 5.25
521.5 RPM

B Dia. 5.375

C Dia. 2.625

drive belts

A Dia 3.25
1725 RPM

Example- Find the RPM of Pulley D

$$\frac{A \times C}{B \times D} \times \frac{\text{RPM of Drive}}{\text{Pulley A}} = \frac{\text{RPM of Drive}}{\text{Pulley D}}$$

$$\frac{3.25 \times 2.625}{5.375 \times 5.25} \times 1725 = \frac{\text{Rpm of Drive}}{\text{Pulley D}}$$

$$\frac{8.53125}{28.21875} \times 1725 = \frac{\text{Rpm of Drive}}{\text{Pulley D}}$$

$$0.3023256 \times 1725 = 521.5 \text{ RPM}$$

Example- Find the RPM of Pulley A

$$\frac{\text{RPM of Pulley D}}{\left(\dfrac{\text{Dia. of A} \times \text{Dia. of C}}{\text{Dia. of B} \times \text{Dia. of D}}\right)} = \text{RPM of Pulley A}$$

$$\frac{521.5}{\left(\dfrac{3.25 \times 2.625}{5.375 \times 5.25}\right)} = \left(\dfrac{521.5}{\dfrac{8.53}{28.21875}}\right)$$

$$\frac{521.5}{0.3022812} = 1725.2 \text{ RPM}$$

Compound Drive System *(continued)*

Finding the diameter of the intermediate pulleys.

Example- Using the previous pulley system, what
would be the size of Pulleys B and C if they
were unknown?

You know that Pulley C will have to be smaller
than Pulley A. Make Pulley C 2.625 inches
Now solve for Pulley B.

$$\dfrac{\left(\dfrac{\text{Pulley A Dia. X Pulley C Dia.}}{\text{Pulley D Dia.}}\right)}{\left(\dfrac{\text{Pulley D rpm}}{\text{Pulley A rpm}}\right)} \qquad \dfrac{\left(\dfrac{3.25 \text{ X } 2.625}{5.25}\right)}{\left(\dfrac{521.5}{1725}\right)}$$

$$\dfrac{\left(\dfrac{8.53125}{5.25}\right)}{0.3023188} = \dfrac{1.625}{0.3023188} = 5.375 \text{ Diameter} \atop \text{of Pulley B}$$

Mechanical

Gear Ratios

$$\text{Gear Ratio} = \frac{\text{Number of teeth on drive gear}}{\text{Number of teeth on driven gear}}$$

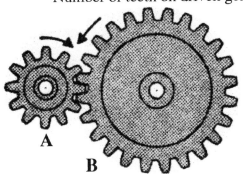

A

B

Example- Using the above example drive Gear A has 12 teeth and gear B has 24 teeth. What is the gear ratio.

$$\text{Gear Ratio} = \frac{24}{12} = 2 \text{ to } 1 \text{ or also stated as } 2\!:\!1$$

Example- If gear B was the drive gear and gear A was the driven gear, the ratio would be-

$$\text{Gear Ratio} = \frac{12}{24} = 0.5 \text{ to } 1 \text{ or } 0.5\!:\!1$$

Example- If the drive gear A had 17 teeth and gear B had 49 teeth the ratio would be-

$$\text{Gear Ratio} = \frac{49}{17} = 2.8824 \text{ to } 1 \text{ or } 2.8824\!:\!1$$

Torque

Torque is usually stated in foot pounds. 1 foot pound is 1 pound of pressure exerted 12 inches from the center of the item being torqued.

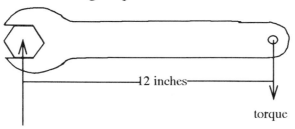

Center of item being torqued

Example - In the above example, if 80 pounds of pull or pressure was put on the wrench, the bolt would have been tightened at 80 foot pounds.

Example - If the wrench in the above example was 18 inches long the amount or torque would have been -

18 divided by 12 = 1.5
1.5 X the amount of pull 80 = 120 foot lb..

Mechanical

Centrifugal Force

If Speed is in Miles Per Hour

$$CF = K \times Speed \times \frac{Speed}{Radius} \times Weight$$

K = 0.0671 if speed is in Miles Per Hour and the Radius is measured in feet.

K = 0.375 if speed is in Feet Per Second and the radius is measured in inches.

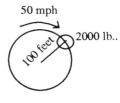

$$CF = 0.0671 \times 50 \times \frac{50}{100} \times 2,000$$

$$CF = 0.0671 \times 50 \times 0.5 \times 2,000$$

$$CF = 3,355 \ lb..$$

If Speed is in Revolutions Per Minute

$$CF = RPM \times RPM \times \frac{Radius}{K} \times Weight \ in \ lb..$$

K = 2,925 if Radius is in feet.

K = 35,100 if Radius is in inches.

$$CF = 1,500 \times 1,500 \times \frac{12}{35,100} \times 4$$

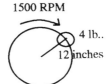

$$CF = 1,500 \times 1,500 \times 0.000342 \times 4$$

$$CF = 3,078 \ lb..$$

Mechanical

Horsepower

One horsepower equals work at the rate of 550 foot pounds per second or 33,000 foot pounds per minute.

$$Hp = \frac{\text{Weight in pounds X Number of feet}}{550 \text{ X Time in Seconds}}$$

Example - A motor weighing 650 pounds is to be lifted 5 feet in 15 seconds. How much horsepower will be used?

$$Hp = \frac{650 \text{ X } 5}{550 \text{ X } 15} = \frac{3250}{8250}$$

$$Hp = 0.3939 \text{ Horsepower}$$

Force

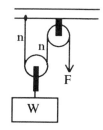

How much force in pounds is needed to move the weight W?

Example - If the weight is 250 lb. how many pounds of pull is needed to raise the weight?

$$F = \frac{1}{n} \text{ X Weight}$$

$$F = \frac{1}{2} \text{ X } 250 = F = 125 \text{ pounds of pull}$$

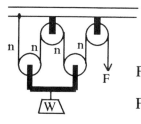

Example - How much pull is needed to raise 1,250 lb..

$$F = \frac{1}{n} \text{ X Weight}$$

$$F = \frac{1}{4} \text{ X } 1,250$$

$$F = 0.25 \text{ X } 1,250$$

$$F = 312.5 \text{ lb.. of pull}$$

Mechanical

Levers

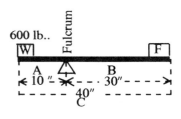

How much weight is needed at point F to lift weight W ?

$$F = \frac{W \times A}{B}$$

$$F = \frac{600 \text{ lb.} \times 10 \text{ inches}}{30 \text{ inches}}$$

$$F = \frac{6000}{30}$$

F = 200 lb.. will equalize the weight W of 600 lb..

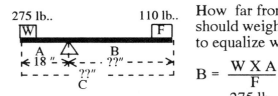

How far from the fulcrum should weight F of 110 lb.. be, to equalize weight W of 275 lb.?

$$B = \frac{W \times A}{F}$$

$$B = \frac{275 \text{ lb.} \times 18}{110 \text{ lb.}}$$

$$B = \frac{4950}{110} \quad B = 45 \text{ inches}$$

$$C = 45 + 18 = 63 \text{ inches}$$

Engine Hoist

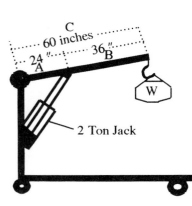

What is the maximum lift capacity of the example hoist?

$$W = \frac{\text{Jack capacity} \times A}{\text{Distance of arm C}}$$

$$W = \frac{4,000 \text{ lb.} \times 24 \text{ inches}}{60 \text{ inches}}$$

$$W = \frac{96,000}{60} \quad W = 1,600 \text{ lb.}$$

Temperature

	Fahrenheit	Celsius	Kelvin	Rankine	Reaumur
Water Boils	212^0	100^0	$373.15K$	671.67^0R	80^0Re
Water Freezes	32^0	0^0	$273.15K$	491.67^0R	0^0Re
Absolute Zero	-459.67^0F	-273.15^0C	$0 K$	0^0R	

Celsius to Fahrenheit

F = (Celsius X 1.8) + 32

Example- change 45^0 Celsius to Fahrenheit

F = (45 X 1.8) + 32   F = 81 + 32 = 113^0 F

Fahrenheit to Celsius

C = (F – 32) X 0.556

Example- change 98^0 Fahrenheit to Celsius

C = (98 – 32) X 0.556   C = 66 X 0.556 = 36.696^0 C

Earth

Temperature (*continued*)

Kelvin Scale uses the same increments as the Celsius Scale.
It starts at absolute zero which is - 273.15 C. It is used
less than the Fahrenheit and Celsius scales. It is used
more by scientists.

Kelvin (K) to Celsius (centigrade)(C)

$$C = K - 273.15$$

Example- change 125 K to Celsius

$$C = 125 K - 273.15 = - 148.15 \,^{\circ} C$$

Celsius (C) to Kelvin (K)

$$K = C + 273.15$$

Example- change $110 \,^{\circ} C$ to Kelvin

$$K = 110 + 273.15 = 383.15 \ K$$

Example- change $- 45 \,^{0} C$ to Kelvin

$$K = -45 + 273.15 = 228.15 \ K$$

Rankine Scale uses the same increments as the Fahrenheit
scale, however its zero is - 459.67 F.

Rankine (R) to Fahrenheit (F)
$$F = R - 459.67$$

Fahrenheit (F) to Rankine (R)
$$R = F + 459.67$$

Rankine (R) to Celsius (C)
$$C = [(R - 459.67) - 32] \ X \ 0.56$$

Example- change 500 R to Celsius

$$C = [(500 - 459.67) - 32] \ X \ 0.56$$
$$C = 40.33 - 32 \ X \ 0.56 = 8.33 \ X \ 0.56 = 4.66 \ C$$

Temperature (*continued*)

Re'aumur Scale is based on a water/alcohol mix. Each
increment in its scale is 4/5 (0.80) of the Celsius scale.

Re'aumur (Re) to Celsius (C)

$$C = 1.25 \ X \ Re$$

Celsius (C) to Re'aumur

$$Re = 0.80 \ X \ C$$

Absolute Zero – 459.67 $^{\circ}$ F is the theoretical point where all
molecular motion stops.

Earth

Hard Freeze

Below 28 Degrees Fahrenheit

Wind Chill Factor

Take the wind speed, multiply it by 1.5, then subtract from the temperature of the thermometer.

Example- The outside temperature is 34 degrees F
The wind is 40 miles per hour.

Wind chill factor = temperature - mph X 1.5

$$WCF = 34^0 - (40 \times 1.5) = 34 - 60 = -26^0$$

Heating and Cooling Degree Days- a unit of

measurement to determine heating and cooling of buildings.

Heat Degree Days

$$65^0 \text{ subtract } \frac{\text{maximum temp} + \text{minimum temp}}{2}$$

Example- It is 72^0 during the day and 48^0 at night.

$$65 \text{ subtract } \frac{72 + 48}{2} = 65 - \frac{120}{2} = 65 - 60 = 5$$

5 Heating Degree Days

Cooling Degree Days

$$\frac{\text{maximum temp.} + \text{minimum temp}}{2} \text{ subtract } 65$$

Example- It is 89^0 during the day and 64^0 at night

$$\frac{89 + 64}{2} \text{ subtract } 65 = \frac{153}{2} - 65 = 76.5 - 65 = 11.5$$

11.5 Cooling Degree Days

Earthquake / Richter Scale

The scale goes from 1 to 9. Each number represents 10 times the previous number. In other words an earthquake of 7 is 10 times the intensity of a quake of 6. A quake of 8 is 100 times a quake of 6.

Richter Number		Increase in Intensity
1		1
2		10
3		100
4		1,000
5		10,000
6		100,000
	6.25	250,000
	6.5	500,000
	6.75	750,000
7		1,000,000
	7.25	2,500,000
	7.5	5,000,000
	7.75	7,500,000
8		10,000,000
	8.25	25,000,000
	8.5	50,000,000
	8.75	75,000,000
9		100,000,000

To estimate how much more intense one earthquake is over another, do the following.

Example- Quake A is 8.9 and Quake B is 6.7

Quake A
8 = 10,000,000   9 X 10,000,000 = 90,000,000
8.9 = 90,000,000 ⟍_____ *comes from the .9 of 8.9*
Quake B
6 = 100,000   7 X 100,000 = 700,000
6.7 = 700,000 ⟍_____ *comes from the .7 of 6.7*

90,000,000 minus 700,000 = 89,300,000
89,300,000 divided by 700,000 = 127.6

Quake A is 127.6 times stronger than Quake B

Earth

P H

Acidity or Alkalinity (bases) of a solution. The neutral being 7 , which is pure water. 1 is the most acidic and 14 the most alkaline. As with the Richter Scale each number represents a ten fold increase. Example- Orange juice is 10 times more acidic than coffee.

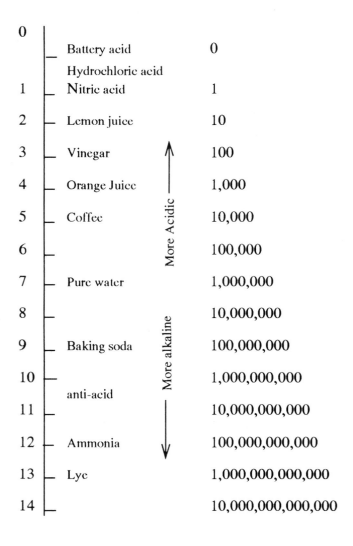

0		
	Battery acid	0
	Hydrochloric acid	
1	Nitric acid	1
2	Lemon juice	10
3	Vinegar	100
4	Orange Juice	1,000
5	Coffee	10,000
6		100,000
7	Pure water	1,000,000
8		10,000,000
9	Baking soda	100,000,000
10		1,000,000,000
	anti-acid	
11		10,000,000,000
12	Ammonia	100,000,000,000
13	Lye	1,000,000,000,000
14		10,000,000,000,000

More Acidic

More alkaline

Sound

Decibels   Units of Energy

Decibels

	Decibels	Units of Energy
Rocket Engine	150	1,000,000,000,000,000
Large Gun,Jet Engine,	140	100,000,000,000,000
Air raid siren up close	130	10,000,000,000,000
Rock Concert, airplane, thunder	120	1,000,000,000,000
Car Horn or alarm	110	100,000,000,000
Subway, helicopter, blowdryer	100	10,000,000,000
Motorcycle, lawn mower, factory	90	1,000,000,000
Garbage disposal, vacuum cleaner	80	100,000,000
City traffic, alarm clock, restaurant	70	10,000,000
Normal conversation, dishwasher	60	1,000,000
Average home quiet street	50	100,000
Leaves in the wind, refrigerator	40	10,000
Whispering, ideal library	30	1,000
Ticking of a watch,	20	100
The slightest sound one can hear	10	10
Threshold of hearing	0	1

For every 10 decibels sound goes up 10 times.
The above scale is basically the same for the Richter Scale and PH.

Example- A Rock Concert is 10 times louder than a Car Horn. This also means a Rock Concert is 1,000,000 louder than normal conversation.

$120 - 60 = 60$ decibels louder (for every 10 decibels sound increases 10 times

$10^6 = 10 \times 10 \times 10 \times 10 \times 10 \times 10 = 1,000,000$

OSHA Noise Limits

Decibels	Maximum Time
120	0
115	15 minutes
110	30 minutes
105	1 hour
100	2 hours
95	4 hours
90	8 hours

Earth

Sound

Adding Noise to Present Environment

When one noise, that is 50 decibels, is added to an environment presently producing 50 decibels of noise, the resulting noise level is not twice as much. To determine the new level of noise do the following formula.

$$\text{Noise Level} = 10 \ X \ \textit{the} \log \textit{of} \left(10^{\frac{\text{decibel}_1}{10}} + 10^{\frac{\text{decibel}_2}{10}} \right)$$

Example- The present level of noise in a shop is 87 decibels. If a new machine is installed that emits 81 decibels, what will the probable noise level be?

$$\text{Noise Level} = 10 \ X \ \textit{the} \log \textit{of} \left(10^{\frac{87}{10}} + 10^{\frac{81}{10}} \right)$$

87 divided by 10 = 8.7   81 divided by 10 = 8.1

$$\text{Noise Level} = 10 \ X \ \textit{the} \log \textit{of} \left(10^{8.7} + 10^{8.1} \right)$$

Calculate 10 $\boxed{Y^x \textit{ key}}$ *8.7* $\boxed{=}$ *5.0119 X 10*[8] *or 501,187,233.6*
Calculate 10 $\boxed{Y^x \textit{ key}}$ *8.1* $\boxed{=}$ *1.2589 X 10*[8] *or 125,892,541.2*

$$\text{Noise Level} = 10 \ X \ \textit{the} \log \textit{of} \ (5.0119 \ X10^8 + 1.2589 \ X \ 10^8 \ )$$

$$\text{Noise Level} = 10 \ X \ \text{the} \log \text{of} \ 6.2708 \ X \ 10^8$$

Calculate 6.2708 $\boxed{\text{EE key}}$ 8 $\boxed{\text{Log key}}$ = 8.79732d

$$\text{Noise Level} = 10 \ X \ 8.7973 = 87.97 \ \text{decibels}$$

Even though the addition of another machine noise at 81 decibels, was less than the previous noise level of 87 decibels, it still increased the over all noise 0.97 decibels more.

Earth

Speed

To change feet per second to miles per hour

60 seconds X 60 =3600   3600 X fps divided by 5,280

Example- Change 1126 feet per-second to miles per
hour.

3600 X 1126 = 4,053,600   divided by 5,280 = 767.7 mph

Sound

Speed of Sound in -

Carbon Dioxide	877	feet per second
Air	1,088	
Hydrogen	4,315	
Water	4,820	
Brass	11,500	
Granite	12,900	
Iron or Steel	16,800	feet per second

Speed of Light - 186,291 feet per second

3600 X 186,291 = 670,650,000 divided by 5280 = 127,016.6
mph

Mach - The ratio of speed as compared to the speed of sound.

Mach = speed of object divided by speed of sound
at 32 degrees and at sea level

Example What is the Mach rating of a jet going 1450 mph.

$$\frac{1450}{742 \text{ mph}} = 1.957 \text{ Mach}$$

Earth

Air Travel - Time Zones

You board a jet plane in New York at 4:00 P.M. and go
to L.A. The distance is 2,571 miles. It takes 6 hours
for a direct flight.
There are 4 time zones of the continent U.S. When you
go west you subtract the number of time zones you
cross, from your watch. In the above example your
watch will read 10 P.M. when you land in L.A.(4 P.M.
plus 6 hours = 10 P.M) However you have crossed 3
time zones, so subtract 3 hours from your watch and set
it for 7:00 P.M. When you go east you add the number
of time zones to your watch.

When you get to the International Time Zone in the
middle of the Pacific Ocean you then add a day to your
clock. You leave New York at 6:00 AM on Sunday and
fly 17 hours to Japan. You will cross 10 time zones
and cross the date line.
6 A.M. plus 17 hours = 11 P.M. on your watch
Subtract the 10 time zones and then you set your watch
at 1 P.M. and because you crossed the International
Date Line it is now Monday.

Speed of Earth

At the equator the Earth is rotating 1,532 feet per second.
The Earth moves around the Sun at 97,786 feet per second

They total 99,318 feet per second -- and you thought
you were standing still.

3600 X 99,318 = 357,540,000 divided by 5280
$$= 67,716 \text{ mph}$$
If the Earth moves around the Sun at 97,786 fps
how many miles does it travel in a year?
3600 X 97,786 = 352,030,000 divided by 5280 = 66,672 mph
24 hours X 365 = 8,760 hours per year
66,672 X 8,760 = 584,050,000 or 5.8405×10^{8}
miles around the sun.

Weights of Things

Metals	Aluminum (cast)	160	lbs per cubic foot
	Gold pure	1204	" "
	Iron (cast)	447	" "
	Lead	711	" "
	Silver	656	" "
	Steel	491	" "

Money   Penny 3.11 grams to 1982, after 2.5 grams
        Nickel   5.0 grams per coin
        Dime   2.5 grams to 1964, after 2.27 grams
        Quarter 6.25 grams to 1974, after 5.75
        Half Dollar 12.5 grams to 1963, after 11.34
        Silver dollar 26.7 grams till 1971 after 24.6

Rock	Granite	165	lbs. per cubic foot
	Sandstone	143	" "
	Limestone	155	" "
	Marble	170	" "
	Slate	183	" "
	Cement (dried)	193	" "
	Concrete (sacked)	137	" "
	Sand (dry)	100	" "
	Sand (wet)	125	" "

Base Rock   3/4 inch rock, dirt and sand
                         cubic yard =   2,000 lbs. dry
                         cubic yard =   2,600 lbs. wet
1/4 inch Rock   cubic yard =   2,400
1 inch Rock   cubic yard =   2,000
Bark   cubic yard =   300 lbs. dry
                         cubic yard =   400 lbs. wet
Earth loose   cubic yard =   2,025
       packed tight   cubic yard =   2,700
Water   62.4 lbs. per cubic foot
                             8.34 lbs. per gallon

Wood Maple   45 lb. per cubic foot
      Pine   32 lb. " " "
      White Oak   50 lb.
      Red Oak   47 lb.
      Popular   32 lb.

Earth-Globe

Longitudes and Latitudes

There are 360 longitudes and they go north and south. Zero longitude goes through Greenwich, England. All longitudes west of Greenwich are designated W and continue west to the International Date Line that is 180 degrees. At 180 degrees, it like 0 degrees is neither west or east. All longitudes east of Greenwich are designated E and go to 180 degrees longitude.

There are 180 latitudes. Zero latitude is the Equator. 90 degrees north is the north pole and 90 degrees south is the south pole.

Each latitude is the same distance, 69.047 miles, from each other.

Each longitude at the equator is 69.1667 miles apart. As they go farther north or south to the poles they get closer together, until they are zero distance apart at 90 degrees north or south.

Distance Between Longitudes

To determine how far longitudes are apart at any given latitude-

      90 - Latitude (sin function) X 69.1667 = miles apart

      Example- How far apart are the longitudes at 32 degrees latitude?

      90 - 32 = 58 (sin function) X 69.1667 = 58.567

      Using a calculator do the following-

90 ⌐ - key⌐ 32 ⌐ = key⌐ ⌐sin key⌐ ⌐X key⌐ 69.1667 ⌐ = ⌐ 58.65668

Earth-Globe

Distance Between Longitudes at listed Latitudes

Latitude	Miles Between Longitudes
O (equator)	69.1667
15	66.81
30	59.90
45	48.91
60	34.58
75	17.90
90	0

Each degree(^{0})has 60 minutes (') and each minute has 60 seconds ("). This then means each degree has 3,600 seconds.

Finding Distances from Points on the Globe-

Example- Latitude the Same - Longitude Different

Point A is 30 ^{0} N   105 ^{0} 32 ' 5" W

Point B is 30 ^{0} N 119 ^{0} 38 ' 20" W

Subtract Point B from Point A

$$119^{0}\ 38\ '\ 20"\ W$$
$$105^{0}\ 32\ '\ 5"\ \ W$$
$$\overline{\ 14^{0}\ \ 6\ '\ 15"}$$

B | 30⁰N | A

119⁰ 38' 20"   105⁰ 32' 5"

Then change the answer to a decimal point.
6 divided by 60 = 0.1 ; 15 divided by 3,600 = 0.004167
Now add the, minutes and seconds, decimal values together. 0.1 + 0.004167 = 0.104167

$$14^{0}\ \ 6\ '\ 15"\ =\ 14.104167^{0}$$

Using the above chart find the distance between longitudes on the latitude of 30 ^{0} which is 59.90 miles.

Multiply 59.90 X14.104167 = 844.8 miles from
                              Point A to B

Earth-Globe

Finding Distances from Points on the Globe-*(continued)*

Example-Latitude the Same - Longitude Different

Point A is $34.5\,^0$N $\quad 60\,^0\,32\,'$ W

Point B is $34.5\,^0$N $\quad 45\,^0\,13\,'$ E

Add Point B to Point A

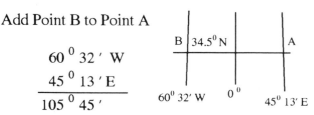

$$60\,^0\,32\,'\text{ W}$$
$$\underline{45\,^0\,13\,'\text{ E}}$$
$$105\,^0\,45\,'$$

Then change the answer to a decimal point.
45 divided by 60 = 0.75 ; Now add the minutes
decimal value to 105 = $105.75\,^0$

Instead of using the chart on page 17 :13 use the formula for distance between longitudes to determine the number of miles each longitude is on the latitude of $34.50\,^0$ N

Remember, at the equator each longitude is 69.16667 miles apart and they get closer together towards the poles.

$$90 - 34.5 = 55.5$$
the sin of 55.5 is 0.82412621 X 69.1667 = 57 miles
57 X 105.75 = 6,027.75 miles

Using a calculator

$90\;\boxed{-}\;34.5\;\boxed{=\text{key}}\;\boxed{\text{sin key}}\;\boxed{\text{X}}\;69.1667\;\boxed{=\text{key}}\;\boxed{\text{X}}\;105.75\;\boxed{=}$

6,027.97 miles

Example   Point A is $32.5\,^0$ N $\quad 105\,^0$ W

Point B is $45.3\,^0$ N $\quad 105\,^0$ W

Subtract Point A from Point B $45.3 - 32.5 = 12.8^0$

Multiply 69.047 X 12.8 = 883.80 miles

Earth-Globe

Finding Distances from Points on the Globe-*(continued)*

Example - Latitude and Longitude Both Different

Point A is 32^0 N and 98^0 W

Point B is 38^0 N and 90^0 W

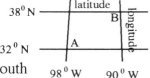

Step 1 Find the difference in North/South

$$38 - 32 = 6^0$$

Step 2 Find the difference in West/East

$$98 - 90 = 8^0$$

Step 3 Change the degrees to miles. Going north or south the space between the latitudes remains the same.

$$6^0 \text{ X } 69.047 = 414.28 \text{ miles}$$

Step 4 The longitudes get closer together as you go from the equator. If you take the longest latitude, it will make up for any difference in the curvature of the earth. So, we will use the 32nd longitude to figure the east/west distance.

$$90 - 32 = 58 \text{ the sin of 58 is } 0.8480481$$

$$0.8480481 \text{ X } 69.1667 = 58.657$$

$$58.657 \text{ X } 8^0 = 469.25 \text{ miles}$$

Step 5 You now know the length of 2 sides of a right triangle, 414.28 and 469.25 Now solve for the hypotenuse.

Distance = $\sqrt{414.28 \text{ X } 414.28 + 469.25 \text{ X } 469.25}$

Distance = $\sqrt{171,627.92 + 220,195.56}$

Distance = $\sqrt{391,823.48}$ = 625.96 miles between Points A and B

Calculate 391,823.48 $\boxed{\text{inverse or 2nd Fctn key}}$ $\boxed{Y^x \text{ key}}$ *2* $\boxed{=}$ *625.96*

Earth

	2000	1990	1900
Population U.S.A.	281,421,906	248,709,873	76,212,168
per square mile	79.6	70.3	21.5
male	138,054,000	121,239,000	38,816,000
female	143,368,000	127,470,000	37,178,000
average age	35.3	32.9	22.9

Households
Husband/Wife families	54,493,232
Families no husband	12,900,103
Living Alone	27,230,075
Living alone over 65	9,722,857
Other than above	1,133,834
Total	105,480,101
Average people per household	2.59

World Population   2000   6.2 billion   1900 1.5 billion

Cars Registered in U.S.A.
	1999	132,432,044
	1995	128,386,775
	1990	133,700,497
	1980	121,600,843
	1950	40,339,077

Cars produced in the U.S.A. year 2000   5,542,519
Trucks produced in the U.S.A. year 2000 7,312,066
Imported Cars   (not trucks) to the U.S.A. 2000   6,324,284
                                      1990   3,944,602
                                      1980   3,116,448
                                      1970   2,013,420
Total Cars produced in the World 2000   42,246,535
   "   Trucks   "   "   "   "   17,519,078

Income/Education   Full time year around employment
	Male	Female
Less than 9th grade	$ 20,787	$ 15,797
up to 12th grade no diploma	25,094	17,918
High School Graduate	34,302	24,967
Some college, no degree	40,339	28,695
Associates Degree	41,948	31,069
Bachelors Degree	56,334	40,413
Masters Degree	68,309	50,139
Professional Degree	99,435	58,978
Doctorate Degree	80,256	57,078

(the above statistics are from U.S. Census Bureau)
30 years of work after a bachelors degree 30X 56,334 = $ 1,690,020
30 years of work with some college no degree 30X 40,339 = $ 1,210,170
The price of a new Corvette or BMW every 3 years for 30 years   479,850
The price of paying college tuition for 96 years at $ 5,000 per year

Curtains

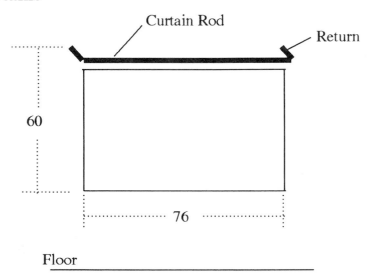

Curtain Rod

Return

60

76

Floor

Before we start measuring for curtains it is necessary to understand the general construction of drapes or curtains. Material in a roll is like the grain in wood and it goes length wise in a roll of material.

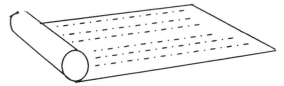

Curtains when made correctly have to hang with the material grain going up and down.

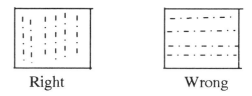

Right   Wrong

If the curtains are not cut and hung correctly they will sag. Also when cutting and ordering the amount of material you have to take into consideration the repeat of a pattern.

Sewing

Curtains (continued)
Measuring for Amount of Fabric

Width - Measure the area you want the curtains to cover. This area maybe beyond the area of the window. To this measurement add for both returns of the rod. Curtains should be 2 1/2 to 3 times longer than this measurement.

Example - The width in the example is 76 inches wide. Now add for the returns 76 + 3 + 3 = 82. We want the curtains to be very full so we will multiply 82 X 3 = 246 inches. There is going to be 2 curtains covering one window, so divide 246 by 2 = 123 inches. You will have a 4 inch hem on each end of the curtain so add 8 inches to 123 inches = 131. If the material you have selected comes in a 48 inch wide roll you will have to divide 131 inches by 48 = 2.73 widths, actually 3 widths. Now that we know we need 3 widths add 2 inches to 131 = 133. This additional 2 inches is so that we can hem them together. Now again divide 133 by

48   inches = 2.77 widths. 3 widths will still be enough.

Length - Using the example you want the curtain to come just one inch below the window.   60 + 1 + 1 for heading + 4 for top hem + 3 1/2 for the bottom hem =   69.5 inches. For each curtain you will need 3 widths 69.5 inches long.

Total   3 widths X 69.5 = 208.5 inches
208.5 X 2 curtains = 417 inches
417 inches divided by 36 inches = 11.58 yards

Curtains (continued)
Pleating Curtains

Each curtain, laid out flat, is 133 inches wide and you
have decided to have a pleat about every 4 inches.
Curtains look best with a pleat at the turn of a return (3
inches from the edge next to the wall) and you want the
last pleat to be 2 inches from the inside edge. Now
subtract 5 inches from 133 = 128. These 2 pleats will
take up 12 inches of material, so subtract 12 from
128 = 116. Each pleat (6 inches) and spacing (about 4
inches) will take about 10 inches per pleat.
Now divide 116 by 10 = 11.6 pleats. You can not have
a partial pleat so we will round this off to 11 pleats.
You are going to start with a pleat at 3 inches and end
with a pleat at 2 inches from the inside edge.
11 pleats X 6 inches per pleat = 66 inches.
Subtract 66 from 116 inches = 50 inches for the
spacing. There will be 12 spaces so divide 50 by
12 = 4.166 inches for each space between the pleats.
Rounded off this is about 4 3/16 inch for each spacing,
with one spacing at 4 1/8 inch.

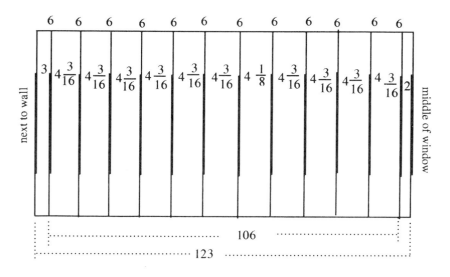

Sewing

Pleated Skirts

Take the waist measurement and add 1 inch of ease for
every 12 inches.
Example waist size 24 + 2 = 26
Multiply X 3   26 X 3 = 78 inches is the width of
    material needed.
Determine distance between pleats. For the example
    we will use 1 1/2 inches.
Divide 78 by 1.5 = 52 pleats.
Subtract waist / hip measurement and ease from width
    of material. 78 - 24 = 54 inches to be used for
    pleats.
Divide 54 by 52 pleats = 1 1/32 inches of material used
    for each pleat.

Full Circle Skirt

Measure the waist then subtract 1 inch.
Example: 27 minus 1 = 26.
Divide 26 by 6 = 4 1/3 inches for the center opening.
To find the amount of material needed, add the skirt
length, top hem, bottom hem and the center opening.
Then double the measurement.
Example: The length is 15 inches, the top hem is 1
    inch, the bottom hem is 1 inch and the
    opening is 4 1/3 inches all equaling 21 1/3
    inches.   Doubling to 42 and 2/3 inches.
The material should be 42 2/3 by 42 2/3 inches square.
Fold the material in half then fold it in half again. You
    now have 4 equal layers.

Draw Radius A then Radius
B. Cut along the lines drawn.
When completed and the
  material laid out flat the
circles should be perfectly
round.

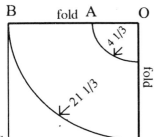

see page 18:6 to understand the above formula

Sewing

Half Circle Skirt

Measure the waist then subtract 1 inch.
Example: 27 - 1 = 26
Divide 26 by 3 = 8 2/3
The Radius A (opening) will be 8 2/3 inches
To find the amount of material needed-
>Add 1 skirt length + waist seam + the hem at the bottom + Radius A.

Example: 15 + 1 + 1 + 8 2/3 = 25 2/3 inches
Now double the amount of material to 25 2/3 X 51 1/3 inches
Fold the material in half and lay out as illustrated.

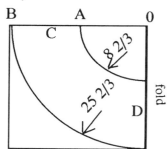

Lay out Radius A 8 2/3
>from Point 0

Layout Radius B 25 2/3
>from point 0

Now sew edges C and D
>together.

Quarter Circle Skirt

Measure the waist then subtract 1 inch.
Example: 27 - 1 = 26
Divide 26 by 1.5 = 17.333 or 17 1/3
The Radius An (opening) will be 17 1/3
To find the amount of material needed-
>Add 1 skirt length + waist seam + the hem at the bottom + Radius A.

Example: 15 + 1 + 1 + 17 1/3 = 34 1/3 inches

The material will have to be 34 1/2 by 34 1/3 inches.

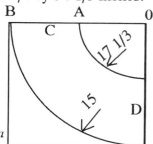

Lay out Radius A 17 1/3
>from Point 0

Layout Radius B 34 1/3
>from point 0

Now sew edges C and D
>together.

see page 18:6 to understand the above formula

Sewing

Why it Works

Example-Full Circle Skirt.

You subtract 1 from the waist measurement of 27 equaling 26. You then divide it by 6, equaling 4 1/3.

Please note that you are making a cut that is the radius of a circle. If you take this radius and use the formula for finding the circumference of a circle you will also be finding the waist measurement (that you already know).

Circumference = 3.1416 X (Radius X 2)

Circumference = 3.1416 X (4 1/3 X 2)

Circumference = 3.1416 X 8 2/3 or 8.66

Circumference = 27.2 inches the original measurement
(let us not count the .2)

Bra Size

Step 1. Measure just below the bust keeping the tape straight across the back. Add 5 inches to this measurement.
Step 2. With the breast properly supported measure around the fullest part of the bust.
Step 3. Subtract Step 1 from Step 2.
Step 4. Check cup size in chart.

Example- Step 1 The tape measures 37 1/2 inches
then add 5 inches equaling 42 1/2
Step 2 The tape measures 45 1/2 inches.
Step 3 Subtract 42 1/2 from 45 1/2 inches = 3
Step 4 The chart shows cup size C.

0	1	2	3	4	5	6
AA	A	B	C	D	DD	DDD or F

Heart Rate While Exercising

The general guide is to take the number 220 and subtract your age. Then multiply by 0.60

For a person 36 years old
(220 - 36) X 0.60
184 X 0.60 = 110 heart beats per minute

The minimum is 50% maximum is 75%.
If the above person has not been exercising they should use the minimum to start off with then slowly work up to the higher rate of 75 %. The rate of 85 % is for extreme exercise and motivation. For a 36 year old the following should be used.

220 - 36 = 184   184 X 0.50 = 92 minimum
                     184 X 0.60 = 110 average
                     184 X 0.75 = 138 high
                     184 X 0.85 = 156 extreme
                     184 X 1.00 = 184 Maximum

Burning Calories

According to most of the charts consulted today, burning calories is proportional to speed, time and weight.

- a person walking 25% faster than they normally do will burn 25% more calories.
- if a person walks twice as far, they will burn twice the number of calories. The same amount the 1st hour as they do the 2nd hour.
- a 200 pound person burns twice as many calories as a 100 pound person.

Health

Burning Calories   **per 10 minutes for every**
100 pounds

Aerobics Average impact	68
Basket ball	64
Bicycling 12-14 mph	64
Bicycling 16-18 mph	96
Bowling	24
Calisthenics	50
Dancing slow	24
Dancing fast, twist,	48
Football (competition)	72
Golf carrying clubs	44
Golf using a chart	28
Jumping Rope	80
Mowing Yard gas engine - push	36
Mowing your in-laws yard	72
Rollerblading	56
Running a 6 minute mile	132
Running a 12 minute mile	64
Setting watching T.V.	9
Skiing cross country	64
Skiing down hill	48
Sleeping	5
Soccer average player	56
Swimming with exertion	80
Swimming butterfly	88
Tennis	56
Walking 3 .3 mph (18 minute mile)	30
Walking 4 mph (15 minute mile)	36
Walking 5 mph (12 minute mile)	45
Water polo	80

Example- A 145 pound woman walks 2 miles in 30 minutes.
How many calories has she burned?

In 1 hour she could walk 4 miles so she is walking at 4
mph and is burning 36 calories per 10 minutes. She
walked for 30 minutes 36 X 3 = 108. She weighs 1.45
more than 100 pounds so multiply 108 by 1.45 = 156.6
calories

Prescriptions for Eye Glasses

Diopters measure the strength of the lens used for glasses or contact lenses. If a prescription starts with a + sign it means the lens is for farsightedness and is convex in shape (thicker in the middle and thinner on the edges.) This means a person can see at a distance but have difficulty seeing up close. If a - sign is used at the beginning of the prescription it means the lens is for nearsightedness and is concave in shape. It is for people who can see good up close but not at a distance.

Astigmatism is due to an irregular cornea (the clear outer surface of the eye). It can be corrected by adding another curvature to the lens in addition to the main curvature, but on a different axis.

A typical prescription for a farsighted person

+ 2.25 - 0.5   axis 48

The above lens is convex with 2.25 diopters and an additional half diopter turned 48 degrees for astigmatism.

A typical prescription for a nearsighted person

- 1.75 + 0.75 axis 27

The higher the first number is the stronger and thicker the lens is. The higher the second number is, means the more astigmatism the person has.

-3.25 + 2.25 axis 95

The above person has so much astigmatism, their cornea so irregular, they would probably not be able to wear soft contact lenses unless they were made for astigmatism (Torex lenses).

Photography
Exposure- f stop

The size of the aperture opening is determined by the focal
length of the lens. This distance is measured from the lens to
the film, where the lens is focused on. The higher the number
the smaller the aperture opening is. The actual area of the
aperture opening doubles with each setting as you go up in
numbers. (f 4 is twice as big as f 5.6)

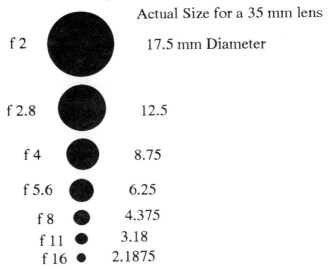

Actual Size for a 35 mm lens

f 2   17.5 mm Diameter

f 2.8   12.5

f 4   8.75

f 5.6   6.25

f 8   4.375

f 11   3.18

f 16   2.1875

Area of the Aperture Opening

To get the diameter, divide the lens size by the f-stop, then
work the problem as area of a circle.

> **Example-** What is the area of the aperture opening of a
> 50 mm lens at f-stop 5.6?

$$\frac{50}{5.6} = 8.929 \text{ mm Diameter}$$

Circle = Radius X Radius X π
8.929 divided by 2 = 4.4645 Radius

4.4645 X 4.4645 X 3.1416 = 62.6176 square mm

Exposure- Shutter speed

The amount of time for a shutter to be open ranges from 1/1000 of a second to 1/15 of a second. Some cameras may not have as much and some may have even a wider range. The typical setting for outdoors is 1/125. Each setting is either double or half as much, depending on which direction you are going. 1/1000 will let in half as much light as 1/500.

Film Speed

ISO (International Organization for Standardization) or ASA (American Standards Association) rates film speed with the same number system. DIN (Deutsche Industrie Norm) uses a different system. Din 24 is the same as ISO 200. Each speed increases by 3 numbers. DIN 27 is ISO 400.

Very slow	50	
Slow	100	outdoors in bright sunlight
Medium	200	flash and outdoors still life
Fast	400	action photos and indoors low light
Extra fast	800	and there are some even faster

Depth of Field

Two elements that influence depth of field is aperture opening and distance. The smaller the aperture opening, the greater depth of field. Using a 50 mm lense, the largest aperture opening f2.8 will give a depth of field of 1 foot, when focused 7 feet away from the camera. At f16 you will get a 10 foot depth of field. The depth of field is more than doubled if focused 14 feet away.

Car

Miles Per Gallon of Gas

After you have filled your gas tank to capacity, record the odometer reading. (the row of numbers in the speedometer). Example 57105.7 odometer reading is fifty seven thousand one hundred five and seven tenths miles. At the end of your drive record the odometer reading. Example 57309.2 miles.

$$
\begin{array}{rl}
\text{Ending odometer reading} & 57309.2 \\
\text{subtract beginning odometer reading} & 57105.7 \\
\text{number of miles driven during test} & 203.5
\end{array}
$$

Look at the gas pump and record the number of gallons used. Example 12.6 gallons. Divide the number of gallons used into the number of miles driven.

203.5 divided by 12.6 equals 16.15 miles per gallon of gas

You can check your mileage with more than one tank full of gas. The longer the trip or test the more accuracy you will have. Total the number of gallons used and divide the total into the total miles driven.

Odometer Accuracy

Along the highway you may see an odometer test sign. Check the right hand number (0-9 tenths) of the odometer when you just pass the "0" sign post. When mile post "1" comes check your odometer reading. For example if you start at between 8 and 7 it should read the same at the "1" mile sign

$$
\begin{array}{lcl}
& 8 & \quad\quad 8 \\
\text{beginning} & 7 & \text{end} \;\; 7
\end{array}
$$

If this is what it reads then your odometer is accurate
However if it reads

$$
\begin{array}{lcl}
& 8 & \quad\quad 7 \\
\text{beginning} & 7 & \text{end} \;\; 6
\end{array}
$$
then your odometer is off 10%. In other words you have traveled one mile and your odometer has recorded .90 miles. This could also mean that you are also going 10% faster than what you speedometer is showing. Instead of going 55 miles an hour you could actually be going 60.5 miles per hour.

Car - Engine Size

Engine sizes, today are advertised in cubic centimeters. Just a few years ago, cars with engines made in the U.S. were measured in cubic inches. All foreign made engines have always been measured in cubic centimeters. To change cubic inches to cubic centimeters multiply by 16.39.

> Example - the American made Ford 302 cubic inch engine - 302 X 16.39 = 4,949.79 Cubic centimeters. However they call this a 5 liter engine. (1,000 centimeters = 1 liter).

To change a metric sized engine to inches divide the cubic centimeters by 16.39 to get cubic inches.
Example- the Mercedes-Benz C230 engine of 2.3 liters

> 2,300 divided by 16.39 = 140.3 cubic inches

The size of the engine is determined by the bore and stroke of the piston. The bore is the diameter of the cylinder that the piston travels in. The stroke is the amount that the piston travels.

> Example - the 351 cubic inch engine with 8 cylinders has a bore of 4 inches and a stroke of 3.5 inches.

> To find the volume of the engine divide the bore of 4 inches by 2 to get the radius of the cylinder. You then work the formula for finding the volume of a cylinder.

Volume = Pi X R^2 X Height

> 3.1416 X 2 X 2 X 3.5

> 3.1416 X 4 X 3.5 = 43.98 Square inches

You then multiply the volume of each cylinder by the number of cylinders.
43.98 X 8 = 351.84 or rounded off to 351 cu. in.

Car - Tire Sizes

tread width

Today we have a mixture of a metric
and inch system for measuring tires. Diameter of wheel

side wall

Example P235 75 R14

width in   how high the tire is   inside diameter of the tire
millimeters   from the ground   in inches(also wheel size)
         in percent of width   *(The R means raial)*

The most important number comes after the R, it tells you the
diameter of the wheel and this can not change unless you get
new wheels. The R stands for Radial. Wheel sizes today range
from 13 to 18 inches. The most common is 14 and 15 inches.

The P stands for passenger tire. The next important size comes
after the P. The higher this number is, the wider the tire tread
is. The wheel width determines how wide the tire can be.

The middle number can be from 40 to 80. The most common
is 65,70 and 75. This number is the percentage of the width,
that determines how high the wheel will be from the ground.
This is also called the sidewall height. You take this number as
a percentage of the width. This is generally close, however the
actual measurement will vary.

Example P 195 65 R 14

Multiply .65 X 195 = 126.75 this is the side wall height
in millimeters

Changing any of the three sizes of a tire- wheel size, width or
side wall height, affects the speedometer.
Example, a tire 205 75 R15 has a diameter of 26 inches, a tire
235 75 R15 has a diameter of 28 inches. This is an increase of
2 inches. 2 divided by 26 = 0.077, or 7.7% increase in the
diameter and thus the circumference of the tire. When the
speedometer reads 65 mph the car will actually be going 7.7%
faster, which is 70 mph. Going from 75 to a 70 side wall
height also produces a significant change in speedometer
readings.

The following is a list of factors that **can** enter into the cost of leasing a car. The list also includes buying the car, however that is not the intentions of leasing a car. The only time leasing should be considered is when you want to have a new car every 3 or 4 years; you do not drive a lot and you take good care of the cars you do drive.

Manufactures Suggested Retail Price(MSRP)		28,000.00
Optional Equipment	+	2,300.00
Dealers Destination Charge	+	400.00
Dealers Preparation Charges	+	225.00
Total Cost of the Car		30,925.00
The Actual Price the Leasing Co. will pay		29,200.00
Down Payment	–	2,500.00
Total Amount of Your Lease (Capitalized Cost)		26,700.00
Residual Value Capitalized Cost X 60%, $26,700 X .60		16,020.00
Depreciation (Capitalized Cost minus Residual Value)		
Example 26,700 minus 16,020		1 0,680.00
Money Factor X (Capitalized Cost + Residual Value)		
X number of payments		
Example .0029 X (26,700 plus 16,020) X 36 months		4,459.97
Lease Acquisition Fee	+	350.00
Total		31,509.97
Monthly payment 31,509.97 divided by 48 equals		656.46
Full Coverage Collision and Liability Insurance		
usually not included in the lease payment per year	+	1,062.00
Refundable Security Deposit		(1,000.00)
Extended Warranty for the Car	+	1,200.00
Vehicle Maintenance Agreement	+	500.00
Gap Insurance	+	350.00
Life and Disability Insurance for 4 years	+	840.00
Sales Tax, Luxury Tax, Title and License	+	2,500.00
usually not included in the lease payment		
Excess Mileage Charge 11-25 cents per milefor miles		
per year above 12 to 18,000 Example 1,350 X 0.15 =	+	202.50
Excess Wear And Tear Example - a tear on seat cover	+	350.00
Refundable Security Deposit.	–	1,000.00
Buy out (residual value)	+	16,020.00
Sales Tax, Smog Inspection, Title Transfer	+	1,375.00
Purchase Option Fee.	+	250.00

Auto - Leasing

Money Factor-This is stated as a decimal, such as 0.00345
It does not relate to an annual interest rate. There is
not an **APR** (annual percentage rate) when it comes to
leasing. To apply the money factor you first add the
Residual Value of the car to the Capitalized Cost and
then multiply the money factor against that figure.

Example- The Capitalized Cost of a car is $28,000 and
the Residual Value is $12,000 after 3 years.
The Money Factor is 0.00345.

$ 28,000 + 12,000 = $ 40,000
$ 40,000 X 0.00345 = $138.00 added to the
monthly payment

Please note that in the above example the Residual
Value was added to the Capitalized Cost. This is not a
mistake. This is one of the reasons the Money Factor
can not be relate to an annual percentage rate.

Depreciation- This is a cost that both the lessor and the lessee
bear, however the lessee pays for it.

Example- A car with a Capitalized Value is $25,000
and the residual Value of 35% is $8,750.
If the lease is 36 months what will be added
to the monthly payment, to pay for the
depreciation?

25,000 - 8,750 = $ 16,250
16,250 / 36 = $ 451.39

If the lessee can negotiate an increase of 5% for the
Residual Value he will save more than that percentage
in that portion his payment.

25,000 X 0.40 = $ 10,000
25,000 - 10,000 = $ 15,000
$15,000 / 36 = $ 416.67 which is a difference
of 7.7 percent.

Capitalized Cost- This is the total cost of the car after add on's, taxes, title, license fee, and insurance.

Gap Insurance- This insures you against a loss if the car is totaled, before you have built up enough equity in the form of lease payments.

> **Example-** You have made 2 monthly lease payments and then you get in an accident that totals the car. The insurance company pays for the car but only for the resale value of a used car. This will be less than the amount remaining on the lease, because as soon as you drive the car off the lot it has depreciated more than what you have remaining on the lease. (assuming you did not pay an exorbitant amount as down-payment).

Refundable Security Deposit- The amount of money you receive back from your initial security deposit, is not of the same value as when you paid it, because of inflation.

> **Example-** If you deposit $800.00 today, what will it be worth 3 years from now if inflation is 4% per year?

1.04 $\boxed{Y^x \text{ key}}$ 3 $\boxed{\frac{+}{-} \text{ key}}$ = 0.8889964
0.8889964 X 800   = $711.20 value 3 years from now.

Collision and Liability Insurance- The leasing company will be the owner of the car and will be exposed to any law-suit should the car be in an accident. They want to limit this exposure and will require the driver to have and pay for usually 3 times the minimum required liability insurance for the car.

Paper

Weights of Paper

Book Paper weight is based on 500 sheets 25 X 38 inches. For example 60 lb. book paper means that 500 sheets 25 X 38 inches weights 60 lbs. The typical weights of book paper ranges from 50 to 100 lbs.

Business and Bond Paper is based on 500 sheets 17 X 22. Basic weights are 13, 16, 20 and 24. Ledger paper weights are 24, 28, 32 and 36. Duplicator papers are 16, 20 and 24 lbs.

Cover Paper is based on 20 X 26 paper. Uncoated cover paper basis weights are 50, 65, 80 and 2 ply 65 lb. Coated cover weights are 60, 80 and 100 lb.

Index Paper weight size is 500 sheets of 25 1/2 X 30 1/2. The typical weights are 90 and 110.

Paper Cutting

The object of cutting large sheets down to small sheets is to get the most out of each sheet. The direction of the grain has to be considered and which direction the grain will go. With heavier weights of paper it may not be a factor. Sometimes plotting out the cuts with sheet grain going in different directions can produce a higher yield.

Example- How many sheets of 8 1/2 X 11 can be cut from a sheet 22 1/2 X 28 1/2 ?

With the Grain

$$\frac{22\ 1/2\ \text{X}\ 28\ 1/2}{8\ 1/2\ \text{X}\ \ \ 11}$$ *Change the fractions to decimals*

$$\frac{22.5\ \text{X}\ 28.5}{8.5\ \text{X}\ \ 11}$$

22.5 divided by 8.5 = 2.64
28.5 divided by 11   = 2.59
Total Sheets = 4

Across the Grain

$$\frac{22.5\ \text{X}\ 28.5}{11\ \text{X}\ \ 8.5}$$

22.5 divided by 11 = 2.04
28.5 divided by 8.5 = 3.35
Total Sheets = 6

Cutting across the grain yields the most sheets.

Paper

Making Columns

To divide a sheet into 2 or 4 equal columns can be done by folding the sheet of paper, but not 3 or 5 columns. To divide a sheet of 8 1/2 X 11 into 3 columns, place a ruler at a slant so that 0 is at one edge and 12 is at the other edge. 12 divided by 3 columns = 4 inches. Make a mark at 4 and 8 inches.

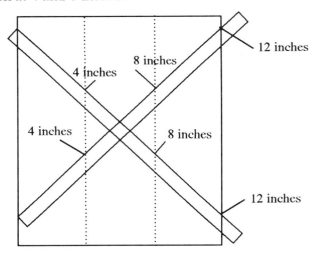

What works even better is to use a Centimeter Ruler Now mark off for 5 equal columns

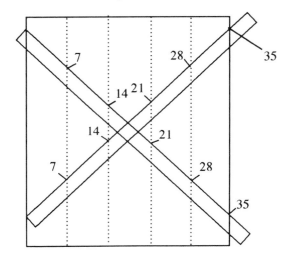

Paper

S1 Metric Size- The international method of sizing and
cutting paper.

Millimeters	Inch Equivalents
A 0 841 X 1189	33.11 X 46.81
A 1 647 X 841	23.39 X 33.11
A 2 420 X 841	16.54 X 23.39
A 3 297 X 420	11.69 X 16.54
A 4 210 X 297	8.27 X 11.69
A 5 148 X 210	5.83 X 8.27
A 6 105 X 148	4.13 X 5.83
A 7 74 X 105	2.91 X 4.13
A 8 52 X 74	2.5 X 2.91

Please note that when cutting a sheet to the next smaller
size there is usually a trim of a half inch.

Copy Size

Reduction or Enlargement

To find the reduction or enlargement you want, divide the size you want by the size you have.

Example- You have an image that is 13 inches long and you want it to fit in an area 10 inches long.

$$\frac{\text{the size you want}}{\text{the size you have}} = \frac{10}{13} = 0.77 = 77 \text{ \% of the original size}$$

Fitting Copy

The amount of reduction/enlargement a camera is set at will do the same percentage change for both the width and length. However, changing the width and length the same amount may not properly fill the required space. This could then require the length or width of the original copy size be changed.

Example- A page of typing is presently 8.5 by 11 inches. More typing can be added to the length but not to the width. The copy has to fit a 7 by 10 inch space.

$$\frac{\text{the size you want}}{\text{the size you have}} \qquad \frac{7}{8.5} = 0.8235 \qquad \frac{10}{11} = 0.9091$$

If you reduce the 11 inches by 0.8235 it will be 9.06 inches and not the required 10 inches. To fit properly the original copy needs to be longer. To find this size do the following.

$$\frac{\text{variable size}}{\text{non variable size}} \qquad \frac{0.9091}{0.8235} = 1.1039$$

Now multiply the length of 11 by 1.1039 = 12.143 inches. The original copy needs to be 8.5 by 12.143 and be reduced by 0.8235 (82.35 %).

Check- 8.5 X 0.8235 =7   12.143 X 0.8235 = 10

Music

Tuning factor-- The standardized scale differences are
1.05946 from one to another.

Frequencies

C	259	
C#	274	274 X 1.05946 = 290.29
D	290	
D#	308	
E	326	326 X 1.05946 = 345.38
F	345	
F#	366	
G	388	
G#	411	
A	435	
A#	461	
B	488	488 divide by 1.05946 equals 460.6

Based on 10 women and 10 men

Reception / Cocktail Party (duration of about 3 hours long)

	Heavy	Moderate	
Soft Drinks			The variety of drinks depends on the people invited and if there is mixed drinks or special wine. The time of day, the type of people and occasion is a factor. If you do not know the people, then you should have double the amount and distributed equally of each type.
Beer			
Liquor	80 total	60 total	
Wine			

Hors D'oeuvres	6 lbs.	4 lbs.	Time of day, people, the occasion and the kind of hors D'oeuvres determines the amount.

Wedding

Champaign	19 bottles 32oz. each	11 bottles 32 oz. each	Figure 6-12 oz. per person every hour

Dinner Party

Wine	10 bottles 32 oz. each	6 bottles 32 oz. each	Figure 6 oz. per serving 1 1/2 to 2 1/2 glasses per person
Appetizers	24	20	

Bread	55 rolls	45 rolls	

Meat	Steak or Fish filet 10 lbs	8 lbs.	If you are going to have 2 types, then decrease each type by 1/4. If steak and chicken then it would be 7 1/2 lbs. of steak and 12 3/4 lbs of chicken
	Hamburger 14 lbs.	10 lbs.	
	Chicken 17 lbs.	13 lbs.	
Vegetable	80 oz.	50 oz.	

Potatoes or Rice or Pasta	120 oz.	80 oz.	

Salad	equivalent heads of lettuce	3 heads	2 heads	Naturally there are many other items for a salad, and these can be added as one wishes.

Desert

Ice-cream	4 quarts	3 quarts	The amount of deserts are really dependent on who the people are and time of day.
Pie or Cake	24 pieces	16 pieces	

(*The richer the desert, the smaller pieces will be used.*
If sheet cake is served larger pieces will be used.)

If your in-laws are invited, double the amount for the meat, wine, beer and liquor.

Notes

Index

A

addition
   signed numbers 4:1,2
absolute zero 17:3
accounts receivable 8:7
accounts receivable to sales year
   ratio 8:15
accounts receivable turnover
   ratio 8:15
acid test 8:14
acre measuring for 15:4
acre feet of water 15:5
acute angle 5:15
acute triangle 5:1,18
adding random numbers 3:3
addition 3:1-2
   decimals 3:17
age of inventory ratio 8:15
air travel 17:10
algebra 4:1-14
   factoring 4:10-13
   foil 4:8
   multiplying terms 4:6
   missing values 4:5,6
   order of operations 4:3
   polynomials 4:7-13
   signed numbers 4:1,2
   simplifying expressions 4:3,4
alternating current 13:4-6
Alternative Depreciation System
   (ADS) 8:9
ammonia 17:6
ampere 12:1
AND logic 4:13
angles 5:15,16,18,20-32
   cutting angles 14:1,13
annual percentage rate (APR) 9:4-7
annuity - ordinary 9:24-37
annuity due 9:24-37
annuity programs 9:24-37
anti-acid 17:6
aperture opening 19:4
appliances 13:4
APR 9:4-7,12

auditorium size 7:1
automobile 19:6-11
automobile leasing 19:9-11
arc 5:4,14
area 1:19, 2:1, 5:3-8
   U.S.A. 17:16
area of a roof 14:15-20
astronomical numbers 2:6
average 6:1
average cost - bonds 8:22
average income bonds 8:21,22
avoirdupois weights 1:20, 2:1

B

baking soda 17:6
ball 5:5,11
barrel 2:2, 2:4
batteries 13:1
battery acid 17:6
beams 14:3-6
belt length 16:1
binary number system 3:30-31
binomial 4:3,7
business ratios 8:14-20
board feet (wood) 14:8
bolt of cloth 12:3
bonds 8:21,22
Boolean Algebra 4:14
bra size 18:16
break even point 8:4
bricks 14:9
British liquid 2:1
British Thermo Unit (Btu) 12:2
bushel 1:20, 2:2

C

cable 2:3
calories 19:1,2
capitalized cost in leasing 19:11
carpet measuring 12:6

Index

Index

Index

Index

Index

Index